AF352852

Advances in Chemical Vapor Deposition

Advances in Chemical Vapor Deposition

Editor

Dimitra Vernardou

MDPI • Basel • Beijing • Wuhan • Barcelona • Belgrade • Manchester • Tokyo • Cluj • Tianjin

Editor
Dimitra Vernardou
Hellenic Mediterranean
University
Greece

Editorial Office
MDPI
St. Alban-Anlage 66
4052 Basel, Switzerland

This is a reprint of articles from the Special Issue published online in the open access journal *Materials* (ISSN 1996-1944) (available at: https://www.mdpi.com/journal/materials/special_issues/Chemical_Vapor_Deposition).

For citation purposes, cite each article independently as indicated on the article page online and as indicated below:

LastName, A.A.; LastName, B.B.; LastName, C.C. Article Title. *Journal Name* **Year**, *Volume Number*, Page Range.

ISBN 978-3-03943-923-2 (Hbk)
ISBN 978-3-03943-924-9 (PDF)

Contents

About the Editor . . . . **vii**

Dimitra Vernardou
Special Issue: Advances in Chemical Vapor Deposition
Reprinted from: *Materials* **2020**, *13*, 4167, doi:10.3390/ma13184167 . . . . **1**

Rukmini Gorthy, Susan Krumdieck and Catherine Bishop
Process-Induced Nanostructures on Anatase Single Crystals via Pulsed-Pressure MOCVD
Reprinted from: *Materials* **2020**, *13*, 1668, doi:10.3390/ma13071668 . . . . **5**

**Huzhong Zhang, Detian Li, Peter Wurz, Yongjun Cheng, Yongjun Wang, Chengxiang Wang,
Jian Sun, Gang Li and Rico Georgio Fausch**
Residual Gas Adsorption and Desorption in the Field Emission of Titanium-Coated
Carbon Nanotubes
Reprinted from: *Materials* **2019**, *12*, 2937, doi:10.3390/ma12182937 . . . . **23**

Xueming Xia, Alaric Taylor, Yifan Zhao, Stefan Guldin and Chris Blackman
Use of a New Non-Pyrophoric Liquid Aluminum Precursor for Atomic Layer Deposition
Reprinted from: *Materials* **2019**, *12*, 1429, doi:10.3390/ma12091429 . . . . **35**

Xia Liu, Lianzhen Cao, Zhen Guo, Yingde Li, Weibo Gao and Lianqun Zhou
A Review of Perovskite Photovoltaic Materials' Synthesis and Applications via Chemical Vapor
Deposition Method
Reprinted from: *Materials* **2019**, *12*, 3304, doi:10.3390/ma12203304 . . . . **49**

Charalampos Drosos and Dimitra Vernardou
Advancements, Challenges and Prospects of Chemical Vapour Pressure at Atmospheric
Pressure on Vanadium Dioxide Structures
Reprinted from: *Materials* **2018**, *11*, 384, doi:10.3390/ma11030384 . . . . **67**

**Kyriakos Mouratis, Valentin Tudose, Cosmin Romanitan, Cristina Pachiu, Oana Tutunaru,
Mirela Suchea, Stelios Couris, Dimitra Vernardou and Koudoumas Emmanouel**
Electrochromic Performance of V_2O_5 Thin Films Grown by Spray Pyrolysis
Reprinted from: *Materials* **2020**, *13*, 3859, doi:10.3390/ma13173859 . . . . **75**

About the Editor

Dimitra Vernardou received her Ph.D. in Physical Chemistry from the University of Salford in 2005. Her thesis was entitled "The Growth of Thermochromic Vanadium Dioxide Films by Chemical Vapour Deposition". During her Ph.D., she designed, optimized and demonstrated an atmospheric pressure chemical vapour deposition (APCVD) reactor for the growth of VO2 and V2O5 coatings. Following this, she undertook postdoctoral studies in IESL-FORTH before being a research fellow in the Centre of Materials Technology and Photonics-TEI of Crete (CEMATEP) in 2006. As a research fellow, she designed and integrated an APCVD reactor for the growth of VO2 (pure or doped with tungsten), V2O5 and WO3 as coatings for thermochromic and electrochromic applications and lithium-ion batteries. Additionally, she has experience in the structural, optical and morphological analysis of coatings. Regarding the electrochemical performance of the metal oxides as electrodes, she has carried out extensive studies using cyclic voltammetry to examine the cyclability, specific capacitance, charge and time response. For the last three years, she has continuously worked with electrochemical impedance spectroscopy to investigate the lithium-ion intercalation/deintercalation mechanism at the electrode–electrolyte interface. She has published over 70 papers in peer-reviewed scientific journals with an h-index of 29 and i10-index of 59 (Source: Google Scholar). She has supervised B.Sc., M.Sc. and Ph.D. students in energy- and environmental-related areas. She has been the principal investigator in three completed research projects involving metal oxide growth and characterization for energy and environmental applications. She currently holds the position of Assistant Professor on Materials for Electrical Energy Efficiency and Storage in the Department of Electrical & Computer Engineering of Hellenic Mediterranean University.

 *materials*

Editorial

Special Issue: Advances in Chemical Vapor Deposition

Dimitra Vernardou [1,2]

1   Department of Electrical and Computer Engineering, School of Engineering,
    Hellenic Mediterranean University, 710 04 Heraklion, Greece; dvernardou@hmu.gr; Tel.: +30-2810-379631
2   Center of Materials Technology and Photonics, School of Engineering, Hellenic Mediterranean University,
    710 04 Heraklion, Greece

Received: 13 September 2020; Accepted: 15 September 2020; Published: 19 September 2020

Abstract: Pursuing a scalable production methodology for materials and advancing it from the laboratory to industry is beneficial to novel daily-life applications. From this perspective, chemical vapor deposition (CVD) offers a compromise between efficiency, controllability, tunability and excellent run-to-run repeatability in the coverage of monolayer on substrates. Hence, CVD meets all the requirements for industrialization in basically everything including polymer coatings, metals, water-filtration systems, solar cells and so on. The Special Issue "Advances in Chemical Vapor Deposition" has been dedicated to giving an overview of the latest experimental findings and identifying the growth parameters and characteristics of perovskites, TiO_2, Al_2O_3, VO_2 and V_2O_5 with desired qualities for potentially useful devices.

Keywords: CVD; electrochromism; perovskite photovoltaic materials; TiO_2; Al_2O_3; VO_2; V_2O_5; computational fluid dynamics

---

In a Chemical Vapor Deposition (CVD) process, the reactants are transported to the substrate surface in the form of vapors and gases. Although there are exceptions, the vapor of the reactive compound, usually an easily volatilized liquid or in some cases a solid, would sublime directly and is generally prepared by injection of the liquid into solvent or heated evaporators [1]. The vapor is then transported to the reaction zone by a carrier gas. The unwanted gas phase nucleation (homogeneous reaction) in CVD can be eliminated through high carrier-gas flow rates, minimum temperatures and cold wall reactors [2].

Would it be possible to assemble nanostructures with confined atomic level thickness, high specific surface area and outstanding surface chemical states at large scale and low cost? CVD is compatible with in-line manufacturing processes where material properties can be controlled with great accuracy, varying growth parameters such as temperature, precursor composition and flow rate. There are various CVD technologies including pulsed-pressure metal organic CVD, atmospheric pressure CVD, atomic layer deposition, spray pyrolysis, plasma-enhanced CVD, aerosol-assisted CVD and so on. There are so many variations on CVD technology because there is no possibility of direct control of the basic processes occurring at the deposition surface. Some of the process technologies that influence the materials' basic characteristics and, as a consequence, their potential application, are included in this Special Issue. The review article by Liu et al. [3] reported on the perovskite photovoltaic materials, with an emphasis on their development through CVD to deal with challenges such as stability, repeatability and large area fabrication methods. In this article, one can gain a clear picture of the influence of different CVD technologies and how the experimental parameters can optimize the perovskite materials for the respective devices.

Pulsed-pressure metal organic CVD (PP-MOCVD) can be utilized for the development of low-cost coatings with both macro and micro-scale, three-dimensional features. Films such as TiO_2 can be uniformly deposited with control of the nanostructure dimension and the coating thickness [4]. Towards this direction, Gorthy et al. [5] highlighted the urgent need for anti-microbial coatings due to the pandemic of COVID-19 through the growth of nanostructured TiO_2 onto handles, push-plates and switches in hospitals. The morphology nanocharacteristic is believed to be the key function for photocatalytic activity with enhanced durability.

CVD at atmospheric pressure (APCVD) is a thin film deposition process with typically high deposition rates. It is an attractive method because it was designed to be compatible with industrial requirements (up-scaling at low cost and high process speed) [6]. The optimization of APCVD towards the development of high yield processes can result in the excellent controllability of the materials' stochiometry, isolating different polymorphs of VO_2 [7]. Among the various polymorphs of VO_2, only the monoclinic VO_2 is a typical thermochromic material [7]. In particular, it is known to undergo a reversible metal-to-semiconductor transition associated with a transformation from monoclinic to tetragonal phase at a critical temperature [8]. Therefore, the utilization of a simple, low cost process with up-scalable possibilities for the development of VO_2 coatings in thermochromic windows is a priority. In the review paper of Drosos et al. [1], the progress on experimental procedures for isolating different polymorphs of VO_2 is outlined. Additionally, the importance of understanding and optimizing the behaviour of the materials supported by modelling studies is highlighted. In that way, theory meets practice, whereas cross-check procedures take place in order to establish firm materials with advanced characteristics.

Atomic layer deposition (ALD) is a process based on the gas phase chemical process in a sequential manner. The majority of ALD processes occur at temperatures > 100 °C with an exception of Al_2O_3. In particular, it can be accomplished with a variety of precursors, in relatively short times and at low temperature [9]. Xia et al. [10] reported the potential to grow Al_2O_3 at 200 °C utilizing different Al precursors via ALD. A consistent 0.12 nm/cycle on glass, Si and quartz substrates was demonstrated to give complex nanostructures with conformity, uniformity and good thickness control as a protection layer in photoelectrochemical water splitting.

Spray pyrolysis is a process in which a precursor solution is atomized through a generating apparatus, evaporated in a heated reactor and decomposed on the top of the substrate into particles and thin films [11]. It is proven to be very useful for the preparation and the design of functional and versatile classes of materials at low cost and easy processing. This process can result in materials with enhanced electrochemical performance for electrochromic applications combined in layered and composite forms for higher reflective property, electrochemical stability and faster electrochromic response [12,13]. In Mouratis et al.'s letter [14], a new approach regarding the development of V_2O_5 electrochromic thin films at 250 °C using ammonium metavanadate in water as precursor is shown. The precursor concentration can affect the morphology of the oxides, resulting in a large active surface area suitable for electrochromic applications.

Conflicts of Interest: The author declares no conflict of interest.

References

1. Drosos, C.; Vernardou, D. Advancements, challenges and prospects of chemical vapour pressure at atmospheric pressure on vanadium dioxide structures. *Materials* **2018**, *11*, 384. [CrossRef] [PubMed]
2. Pulker, H.K. *Coatings on Glass*, 2nd ed.; Elsevier: Amsterdam, The Netherlands, 1985; p. 139.
3. Liu, X.; Cao, L.; Guo, Z.; Li, Y.; Gao, W.; Zhou, L. A review of perovskite materials' synthesis and applications via chemical vapor deposition method. *Materials* **2019**, *12*, 3304. [CrossRef] [PubMed]
4. Gardecka, A.J.; Bishop, C.; Lee, D.; Corby, S.; Parkin, I.P.; Kafizas, A.; Krumdieck, S. High efficiency water splitting photoanaodes composed of nano-structured anatase-rutile TiO_2 heterojunctions by pulsed-pressure MOCVD. *Appl. Catal. B Environm.* **2018**, *224*, 904–911. [CrossRef]
5. Gorthy, R.; Kumdieck, S.; Bishop, C. Process-induced nanostructures of anatase single crystals via pulsed-pressure MOCVD. *Materials* **2020**, *13*, 1668. [CrossRef] [PubMed]
6. Malarde, D.; Powell, M.J.; Quesada-Gabrera, R.; Wilson, R.L.; Carmalt, C.J.; Sankar, G.; Parkin, I.P.; Palgrave, R.G. Optimized atmospheric-pressure chemical vapor deposition thermochromic VO_2 thin films for intelligent window applications. *ACS Omega* **2017**, *2*, 1040–1046. [CrossRef] [PubMed]

7.   Piccirillo, C.; Binions, R.; Parkin, I.P. Synthesis and functional properties of vanadium oxides: V_2O_3, VO_2, and V_2O_5 deposited on glass by aerosol-assisted CVD. *Chem. Vap. Depos.* **2007**, *13*, 145–151. [CrossRef]
8.   Vernardou, D.; Louloudakis, D.; Spanakis, E.; Katsarakis, N.; Koudoumas, E. Thermochromic amorphous VO_2 coatings grown by APCVD using a single-precursor. *Sol. Energ. Mater. Sol. C* **2014**, *128*, 36–40. [CrossRef]
9.   Groner, M.D.; Fabreguette, F.H.; Elam, J.W.; George, S.M. Low-temperature Al_2O_3 atomic layer deposition. *Chem. Mater.* **2004**, *16*, 639–645. [CrossRef]
10.   Xia, X.; Taylor, A.; Zhao, Y.; Guldin, S.; Blackman, C. Use of a new non-pyrophoric liquid aluminium precursor for atomic layer deposition. *Materials* **2019**, *12*, 1429. [CrossRef] [PubMed]
11.   Jung, D.S.; Ko, Y.N.; Kang, Y.C.; Park, S.B. Recent progress in electrode materials produced by spray pyrolysis for next-generation lithium ion batteries. *Adv. Powder Technol.* **2014**, *25*, 18–31. [CrossRef]
12.   Ramadhan, Z.R.; Yun, C.; Park, B.-I.; Yu, S.; Kang, M.H.; Kim, S.K.; Lim, D.; Choi, B.H.; Han, J.W.; Kim, Y.H. High performance electrochromic devices based on WO_3-TiO_2 nanoparticles synthesized by flame spray pyrolysis. *Opt. Mater.* **2019**, *89*, 559–562. [CrossRef]
13.   Mujawar, S.; Dhale, B.; Patil, P.S. Electrochromic properties of layered Nb_2O_5-WO_3 thin films. *Mater. Today Proc.* **2020**, *23*, 430–436. [CrossRef]
14.   Mouratis, K.; Tudose, V.; Romanitan, C.; Pachiu, C.; Tutunaru, O.; Suchea, M.; Couris, S.; Vernardou, D.; Koudoumas, E. Electrochromic performance of V_2O_5 thin films grown by spray pyrolysis. *Materials* **2020**, *13*, 3859. [CrossRef] [PubMed]

 *materials*

Article

Process-Induced Nanostructures on Anatase Single Crystals via Pulsed-Pressure MOCVD

Rukmini Gorthy, Susan Krumdieck * and Catherine Bishop

Department of Mechanical Engineering, College of Engineering, University of Canterbury, 20 Kirkwood Ave, Christchurch 8041, New Zealand; rukmini.gorthy@pg.canterbury.ac.nz (R.G.); catherine.bishop@canterbury.ac.nz (C.B.)
* Correspondence: susan.krumdieck@canterbury.ac.nz

Received: 1 March 2020; Accepted: 1 April 2020; Published: 3 April 2020

Abstract: The recent global pandemic of COVID-19 highlights the urgent need for practical applications of anti-microbial coatings on touch-surfaces. Nanostructured TiO_2 is a promising candidate for the passive reduction of transmission when applied to handles, push-plates and switches in hospitals. Here we report control of the nanostructure dimension of the *mille-feuille* crystal plates in anatase columnar crystals as a function of the coating thickness. This nanoplate thickness is key to achieving the large aspect ratio of surface area to migration path length. TiO_2 solid coatings were prepared by pulsed-pressure metalorganic chemical vapor deposition (pp-MOCVD) under the same deposition temperature and mass flux, with thickness ranging from 1.3–16 μm, by varying the number of precursor pulses. SEM and STEM were used to measure the *mille-feuille* plate width which is believed to be a key functional nano-dimension for photocatalytic activity. Competitive growth produces a larger columnar crystal diameter with thickness. The question is if the nano-dimension also increases with columnar crystal size. We report that the nano-dimension increases with the film thickness, ranging from 17–42 nm. The results of this study can be used to design a coating which has co-optimized thickness for durability and nano-dimension for enhanced photocatalytic properties.

Keywords: anatase single crystals; process-induced nanostructures; competitive growth; pp-MOCVD

1. Introduction

Titanium dioxide has been of high interest for its photocatalytic properties under UV light since the discovery of the Honda–Fujishima effect in 1972 [1]. The best known application of TiO_2 is self-cleaning glass coated with Pilkington Activ[TM] [2]. Anatase and rutile are the most widely researched phases of TiO_2 for photocatalytic applications. The bandgap of anatase is 3.2 eV and the bandgap of rutile is 3.0 eV. Despite the wider bandgap, anatase has high photocatalytic activity (PCA) due to higher surface-adsorption rate of hydroxyl radicals [3]. Anatase also exhibits slower charge recombination rates than rutile [4]. The majority of the studies on TiO_2 photocatalysis investigate a combination of anatase and rutile [5].

TiO_2 nanoparticles have higher specific surface area than bulk titania coatings. Nanoparticles also have a shorter exciton path length from the point of generation to the surface, resulting in lower electron-hole recombination rates for films less than 15 nm [6]. Carbonaceous TiO_2 enhances the PCA in the visible spectrum [7]. Recently, many studies have reported doping TiO_2 with noble metals and other elements to extend the bandgap [8–10].

Nanostructured Materials for Coating Applications

We aim to achieve a nanostructured solid material, which would have the high active surface area and low exciton migration path of nanomaterials, but without the fabrication and handling of nanoparticles. Hashimoto et al. theorized that selective nanostructuring of TiO_2 surfaces would improve the hydrophilicity of the materials thereby making them photocatalytically superior [11]. Nanostructured TiO_2 materials were reported to exhibit improved performance when used as electrodes for lithium-ion batteries compared to electrodes consisting of nanocrystalline anatase [12]. Nanostructured materials such as mesoporous titania have demonstrated improved photoanodic efficiency over Degussa P25 [13]. The main challenge with making practical use of nanostructured materials is in the processing of a solid coating layer on a substrate.

Figure 1 illustrates the differences between nanomaterials and nanostructured materials. Nanoparticles are processed in a hydrothermal solution and are difficult to attach to a substrate. Nano rods and other 2-D structures grown on substrates have not yet been demonstrated for practical coatings. The nanostructured, multiphase, thick solid coating grown by pulsed-pressure metalorganic chemical vapor deposition (pp-MOCVD) has been demonstrated to be adherent and durable [14].

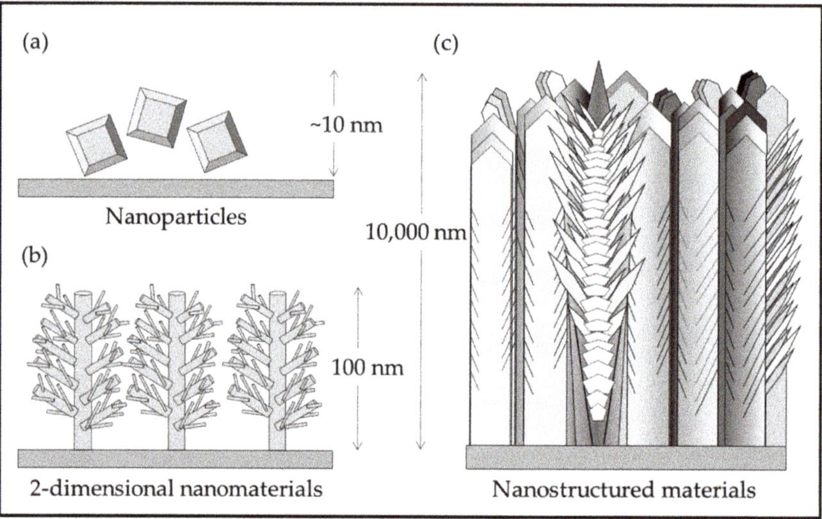

Figure 1. Illustration of (**a**) TiO_2 nanoparticles; (**b**) rutile nanorods with secondary structures grown on a substrate [15]; (**c**) nanostructured TiO_2 in a solid coating adhered to a substrate [14].

Pulsed-pressure metalorganic chemical vapor deposition (pp-MOCVD) has been used to nanoengineer solid TiO_2 coatings (≤17 μm thick) that are composites of anatase, rutile and carbon. The coatings shown in Figure 2 have a rarely-seen columnar morphology composed of single crystal anatase and polycrystalline rutile columns [16]. Anatase columns appear pyramidal at the top and the rough dendritic structures are polycrystalline rutile (Figure 2). The thick robust coatings produced on stainless steel substrates exhibited high antimicrobial activity under visible light [14]. Thinner films on fused silica substrates have formal quantum efficiency 59 times higher than the commercial photocatalyst Pilkington Activ^{TM} [17], measured with stearic acid degradation in UV light [18]. In previous work, we used ASTAR^{TM} analysis of TEM samples of TiO_2 films grown on stainless steel to determine that the anatase columns were single crystals, even though they are made up of nano-plates, with A[220] columnar growth direction [14]. The properties of photocatalytic materials depend on the crystallography and morphology, but there has been no research reporting a study of nanoscale feature size in TiO_2 bulk or coating materials to date.

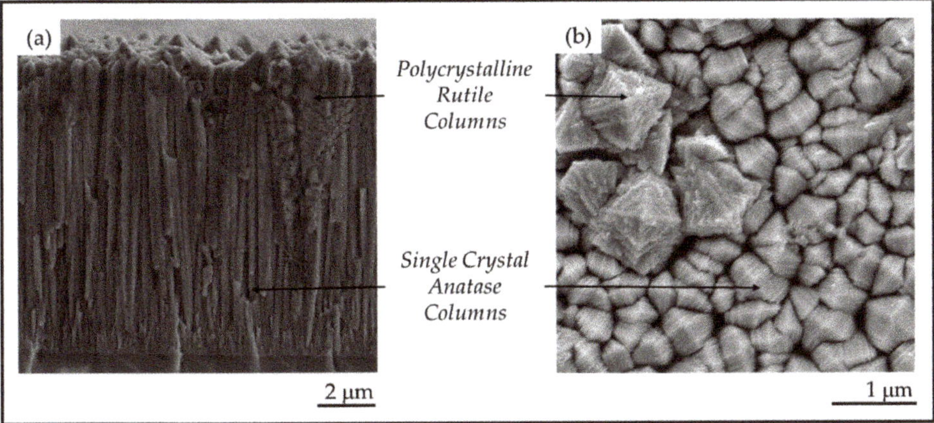

Figure 2. (a) Cross-section morphology on a fracture surface and (b) plan-view surface morphology of TiO$_2$ films prepared by pulsed-pressure metalorganic chemical vapor deposition (pp-MOCVD).

In this study, coatings of TiO$_2$ were produced on glass and stainless-steel substrates. All crystals exhibited the characteristic anatase *mille-feuille* and rutile *strobili* nanostructures shown in Figure 2. The thickness of the plate-like *mille-feuille* structures was observed to depend on the coating thickness, which is controlled by the number of precursor pulses. The thickness of the *mille-feuille* plates in the anatase columns is of great interest because the PCA depends significantly on the nanostructure dimension and the total surface area. The competitive columnar growth of TiO$_2$ anatase single crystals and the observed relation to nanostructure dimensions were investigated.

2. Experimental Details

2.1. Pulsed-Pressure MOCVD Technology

The pp-MOCVD technique is a one-step deposition process that was developed by Krumdieck et al. to coat large objects such as turbine blades with thermal barrier coatings [19]. The process works by direct injection, at timed intervals, of metered volumes of precursor solution via an ultrasonic atomizer into a continuously evacuated deposition chamber. The flash evaporation of atomized liquid in the evacuated reactor chamber produces a sharp pressure spick, and produces a well-mixed reactor condition enabling the coating of complex shaped objects. The objects or substrates are placed on a susceptor that is heated with a water-cooled Cu induction coil. The Titanium tetra-isopropoxide (TTIP) precursor is decomposed when it encounters the heated substrate and forms a macroscopically uniform coating. Solvent and reactant product vapors are evacuated into a liquid N$_2$ cold trap. At high temperatures (>500 °C) the reactor works in the mass-transport controlled regime with high precursor-arrival rate to the substrate surface during the peak of the pressure pulse [20]. The pulsed-pressure cycle is typically 6 seconds with less than 0.5 seconds of pressure rise, and 5 seconds of pump-down [21]. The pp-MOCVD process reduces the reactor and substrate geometry effects on the deposition, making it a versatile technique to coat 3D objects.

2.2. Materials and Chemicals

TiO$_2$ coatings were deposited on 25 mm × 25 mm × 1 mm fused silica substrates (Esco Optics, Oak Ridge, NJ, USA) and a 340 stainless steel substrate (Aperam S.A., Isbergues, France) using a TTIP precursor solution. The precursor is a dilute mixture of 5 mol% of TTIP (>97% Sigma Aldrich, St, Louis, MO, USA) in dry, HPLC-grade toluene with no carrier gas and no additional oxidizing agents.

Table 1 provides a list of samples with their identities (IDs) and the respective number of pulses. All the samples were deposited at 525 °C for fused silica substrates and 500 °C for stainless steel substrates. The material characterizations were carried out on 6 samples prepared on fused silica substrates with the number of pulses ranging from 150 to 1000. Fused silica is snapped to provide SEM analysis of the fracture surface and measurement of the thickness. Sample G was deposited on stainless-steel 304 with 600 pulses and characterized by focused ion beam (FIB).

Table 1. Sample Identifiers, Substrates and Number of Deposition Pulses.

Sample ID	Substrate	Number of Pulses
A	Fused Silica	150
B	Fused Silica	200
C	Fused Silica	350
D	Fused Silica	500
E	Fused Silica	750
F	Fused Silica	1000
G	Stainless Steel	600
SA	Stainless Steel	400
SB	Stainless Steel	735
SC	Stainless Steel	909

2.3. Characterization Methods

The plan-view and fracture surface morphologies of the TiO_2 coatings were observed using a JEOL 7000F Scanning Electron Microscope (SEM, UC, Christchurch, NZ). The samples were scored on the uncoated side using a diamond-tipped scribe and fractured into four sections to expose the cross-section of the coatings. Prior to imaging, the fractured samples were sputtered with Cr using a Quorum Tech rotary pumped coater (UC, Christchurch, NZ). Ten measurements of the film thicknesses were obtained and the mean film thickness (w) reported. The mean growth-rate (GR) of a coating prepared with N pulses was calculated using

$$GR = \frac{w}{N} \tag{1}$$

The mean anatase column diameter was determined from the plan-view SEM images using the ASTM E112 standard circle-intercept method [22]. Five test circles were used to obtain the column size for each sample.

The phase composition of the coatings was determined using a Rigaku SmartLab X-ray diffractometer (XRD, UC, Christchurch, NZ) equipped with a Cu $K_\alpha(\lambda = 1.5148$ Å) source. The as-deposited samples were mounted on a flat sample holder and the detector was set up to collect from 5° to 90° in 2θ at a scan rate of 10° per minute. The spectral peaks were indexed using the RRUFF database [23].

The chemical composition of the TiO_2 samples was analyzed using Surface Enhanced Raman Spectroscopy (SERS. The Raman spectra were obtained using a HORIBA Jobin-Yvon LabRam spectrometer (MacDiarmid Inst./GNS, Wellington, NZ) equipped with an Ar ion (514 nm) laser. The power of the laser was set at 420 μW and the sample surfaces were analyzed as-deposited. The spectra were deconvoluted using a Gaussian peak-fitting algorithm in Origin Pro software (OriginLab, Northampton, MA, USA).

The cross section of sample G was studied using a JEOL 300CF Scanning Transmission Electron Microscope (STEM, IMRI, UC-Irvine, Irvine, CA, USA). The STEM is equipped with a 300 kV cold field emission gun and has a resolution of 80 pm. The sample for STEM imaging was prepared using a FEI Quanta 3D focused ion beam (FIB)-SEM (IMRI, UC-Irvine, Irvine, CA, USA).

2.4. Nanostructure Dimension Measurement

The anatase columns are composed of nanoscale plates. We used a straight-forward measurement technique using the plan-view SEM images to determine the thickness of the plates. Fifteen crystals were selected from the SEM image that were symmetrical and had the most plates visible. The central plates at the peak of the crystals are always slightly thicker, so 3–5 plates as shown in Figure 3 were measured using the GMS Digital Micrograph Suite [24].

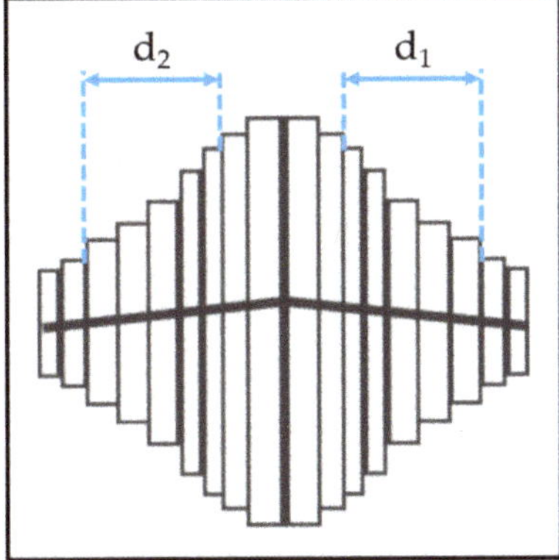

Figure 3. Methods to determine the nanoplate thickness for anatase columns.

Two measurements were taken for each crystal column as shown in Figure 3. If 'n' plates are measured on each side of the central plates, then the thickness of a single nanoplate 't' is calculated as

$$t = \frac{1}{2n}(d_1 + d_2) \tag{2}$$

The measurements are statistically analyzed to obtain the thickness of the nanoscale plates on the anatase columns and the measurement accuracy. This measurement is referred to as the nano-dimension.

3. Results

3.1. Phase and Composition of TiO$_2$ Coatings Prepared by pp-MOCVD

The XRD analysis of the coatings showed that all the coatings were TiO$_2$ with both anatase and rutile phases. The lattice parameters are consistent with stoichiometric TiO$_2$ [25,26]. The XRD pattern given in Figure 4a is representative of all the measurements for the samples in this study. The pattern also shows that the anatase phase exhibits a strong [220] growth texture. This is consistent with previous studies of pp-MOCVD films [14,18,20,27].

A Raman spectrum representative of all the samples in this study is provided in Figure 4b. The deconvoluted peaks correspond to the anatase phase of TiO$_2$ and amorphous carbon present in the films [23,28,29]. The amorphous C peak D$_1$ represents aromatic carbon rings and the peak G$_1$ represents the sp^2 C=C bonds. No evidence for Ti-C was detected by Raman spectroscopy and no bulk titanium carbide phases were identified in XRD. These results are consistent with our previous work, where a uniform distribution of carbon was measured through the films thickness [27].

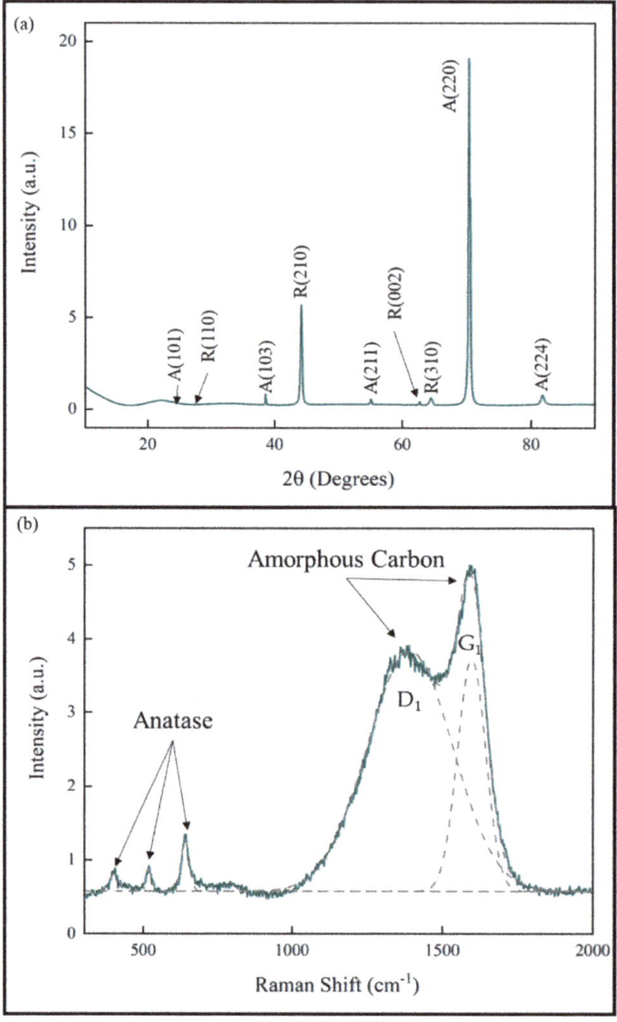

Figure 4. (a) XRD pattern and (b) Raman spectrum of TiO$_2$ coatings prepared by pp-MOCVD.

3.2. Plan-View Surface Morphologies of TiO$_2$ Coatings

The TiO$_2$ coatings were all macroscopically uniform, adherent and black in color. The SEM images of the coatings showed that the coatings were composed of columnar crystals with two distinct morphologies, anatase *mille-feuille* and rutile *strobili* as described in previous work [14]. The top of an individual anatase column resembles a plated pyramid-like structure. Figures 5 and 6 show the plan-view surface SEMs of the samples deposited on fused silica and stainless steel substrates. The figures show that the surface morphologies for TiO$_2$ coatings on stainless steel and fused silica substrates are similar.

The images show that as the number of pulses, and thus the thickness, increases, the number of nanoscale plates and the column diameter also increase. The thickness of the plates is not obviously different, but the measurements reveal a definite trend in nanoscale dimension. The column diameters and the surface-nanostructure dimensions for samples A-F are provided in Table 2.

Figure 5. Plan-view SEM images showing the surface of anatase crystals for TiO_2 samples on fused silica prepared with (**A**). 150 pulses; (**B**). 200 pulses; (**C**). 350 pulses; (**D**). 500 pulses; (**E**). 750 pulses and (**F**). 1000 pulses.

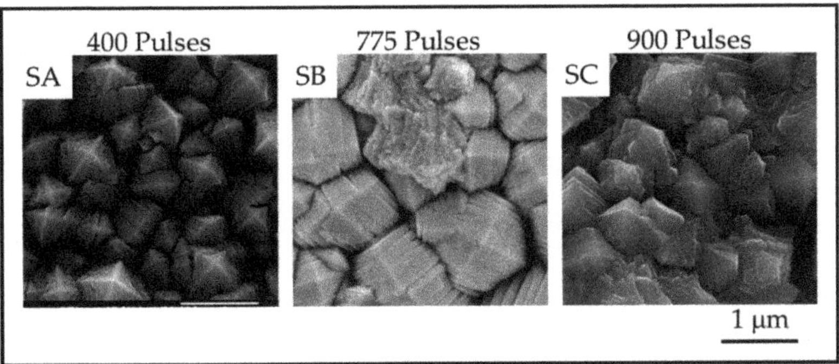

Figure 6. Plan-View SEM images showing the surface of anatase crystals for TiO_2 samples on stainless steel prepared with SA. 400 pulses; SB. 735 pulses and SC. 909 pulses.

Table 2. Column diameters and nanoplate thicknesses reported as mean and standard deviation of the measurements.

ID	Number of Pulses	Column Diameter (nm)	Plate Thickness (nm)
A	150	157 ± 35	17.85 ± 3.5
B	200	257 ± 5	27.98 ± 5.3
C	350	277 ± 5	27.53 ± 2.3
D	500	469 ± 30	36.23 ± 2.7
E	750	553 ± 10	43.25 ± 3.5
F	1000	600 ± 18	43.74 ± 3.0

3.3. Analysis of Growth and Nanoscale Dimensions

Figure 7 shows the variation of the column diameters with number of pulses. It is observed that the column diameter increases with increasing numbers of pulses. The relationship appears to have two linear regions with a break point around 600 pulses. The linear curve fit for all the data shows that the R^2-value is 0.89. The relation at the low number of pulses indicates that the column size of ~100 nm would not have nanoplates, and this is consistent with previous results [18]. The column diameters appear to approach a plateau in the thicker films with higher number of pulses. This behavior fits the model of competitive crystal growth [30].

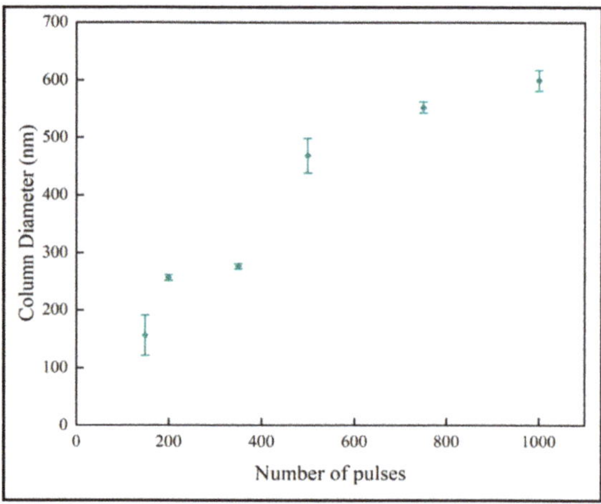

Figure 7. Variation of column diameter with number of pulses.

The nanoplate thickness increases with column diameter as shown in Figure 8. We observe that there is likely a change in the relationship in the later stage as well. This is consistent with the behavior of the samples observed in Figure 7. However, this analysis depends on two difficult measurements which increases the uncertainty, and there is a wide variation in crystal diameter.

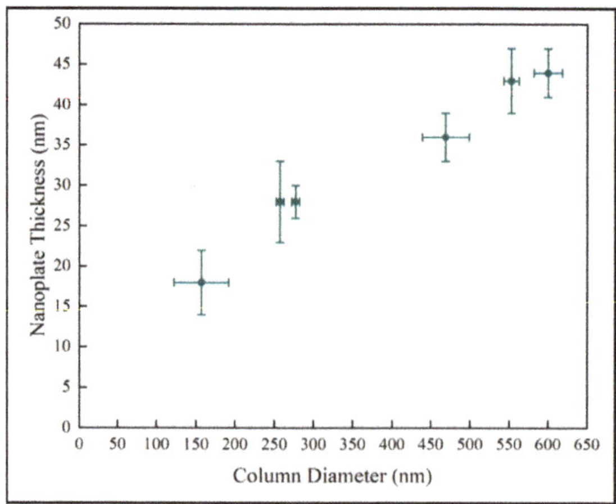

Figure 8. Variation of surface nanoplate thickness with column diameter.

3.4. Fracture Surface Morphologies

The fracture surface images of the coatings on fused silica substrates are shown in Figure 9, and the film thicknesses are provided in the Table 3. The fracture surface SEM images show that columnar morphology extends through the entire thickness of the films. The thickness of the coatings increases with increasing numbers of pulses as expected for a mass transport-controlled deposition.

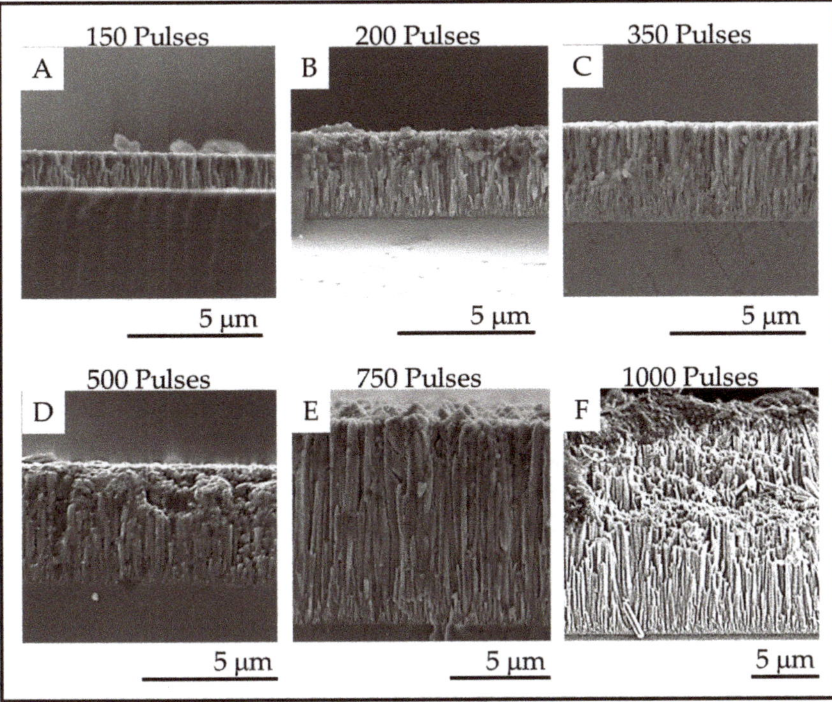

Figure 9. Fracture surface SEM images of the TiO$_2$ coatings deposited on fused silica with (**A**). 150 pulses; (**B**). 200 pulses; (**C**). 350 pulses; (**D**). 500 pulses; (**E**). 750 pulses and (**F**). 1000 pulses.

Table 3. Film thickness and mean growth-rate.

ID	Number of Pulses	Film Thickness (µm)	Growth-Rate (nm/pulse)
A	150	1.33 ± 0.03	8.87 ± 0.2
B	200	3.07 ± 0.05	15.35 ± 0.25
C	350	3.30 ± 0.04	10 ± 0.11
D	500	4.00 ± 0.03	8.4 ± 0.06
E	750	11.50 ± 0.10	15.33 ± 0.13
F	1000	16.03 ± 0.50	16.03 ± 0.5

The variation of the film thickness with number of pulses is shown in Figure 10. The film thickness increases with the number of pulses, which is expected with mass-transport controlled growth. The thinner films below 600 pulses again seem to have a different growth regime than the thicker materials. We do not offer an explanation for this other than observing that the number of nanoplates per columnar crystal also increases dramatically from 500 pulses upward.

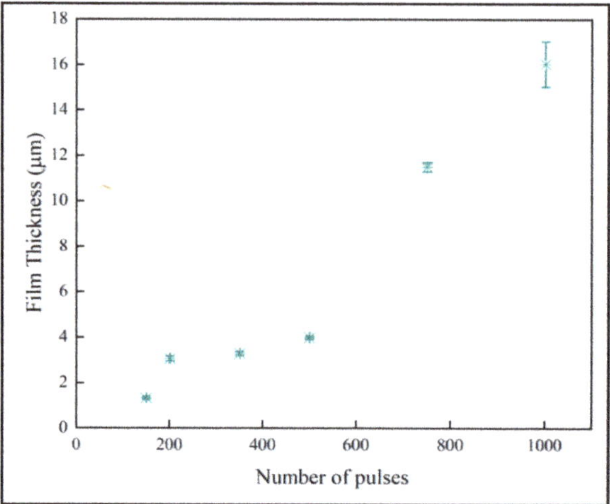

Figure 10. Sample thicknesses increases with number of pulses.

Figure 11 shows the nanoplate thickness rapidly increasing with film thicknesses in the thinner coatings, but then becoming independent of thickness in the thickest films. The result for the thin samples is consistent with a previous study on initial growth of pp-MOCVD films where three growth stages were identified, early, transition and late stage growth [18].

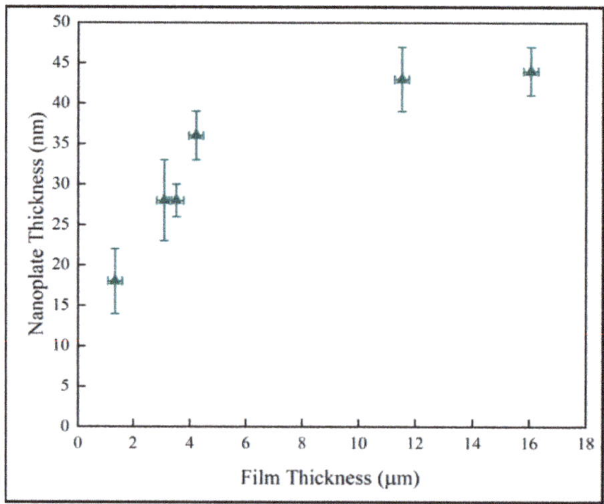

Figure 11. Variation of nanoplate thickness with film thickness.

In Figure 12, the variation of the nanoplate dimension with the mean film growth-rate does not demonstrate any discernable pattern between these quantities. In pp-MOCVD, as in other CVD (Chemical Vapor Deposition) processes, the growth rate is most dependent on deposition temperature. The accuracy of the measurement of temperature and the control of the induction heating means that the variability of the temperature during the majority of a deposition could be +/−15 °C. Given the Arrhenius relationship between growth rate and temperature, the variation of growth rate with pulses could essentially be no relationship, and the data represents processing variability.

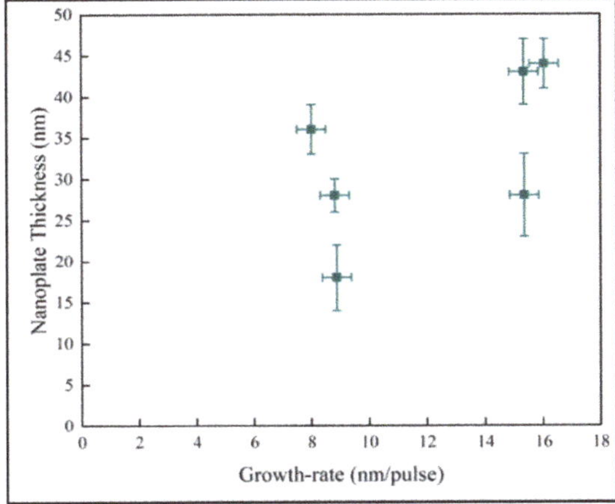

Figure 12. Variation of nanoplate thickness with mean film growth-rate.

3.5. Nanostructure Dimensions along Column Length

Figure 13 shows the STEM image of the cross-section of sample G which is 4.77 μm thick. The white area at the base is the stainless steel substrate. The column width is much smaller at the interface of the film and substrate than at the top of the coating. Regions of Interest (ROI) are indicated which represent roughly 150 pulses of growth each. In ROI-1 near the substrate, it is harder to discern the appearance of nanoplates, and the columns have a small diameter. The columnar crystals have a tapered appearance in regions of interest (ROI) 1 and 2. ROI-2 in particular has clear evidence of the competitive growth resulting in the termination of some columns which are then overshadowed by others. The columns appear straighter in ROIs 3 and 4. This is typical of the Zone-T competitive growth [31].

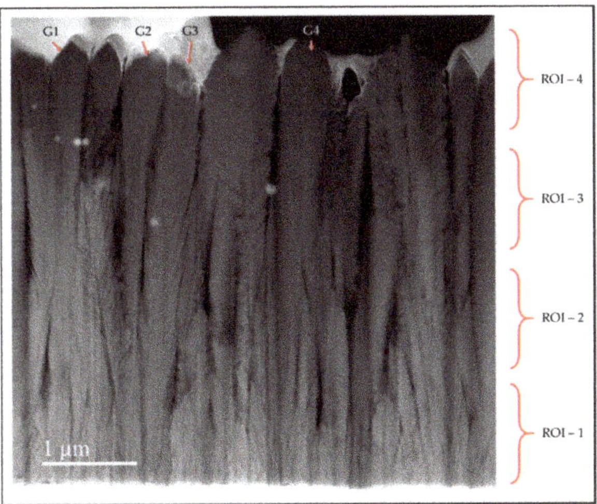

Figure 13. Cross-section STEM image of sample G prepared via pp-MOCVD.

An interesting observation here is the number of crystal columns in each region. There are higher number of crystallites observed in ROI-1 than in ROI-4 where we observe nine distinct crystal columns. This variation of the column diameters from the substrate to the top of the film provides visual evidence of competitive growth in the TiO_2 columnar crystals during a pp-MOCVD process.

The four ROIs of 1 µm length for four anatase columns were analyzed. The anatase columns are indicated on Figure 13 by the labels G1, G2, G3 and G4. The average crystal widths from 10 measurement in each region are provided in Table 4. The column widths of the crystals at the surface of ROI-4 average ~410 nm which is consistent with the ranges of column diameters observed for the samples D and E. At the substrate-film interface in ROI-1, we observe a decrease in the dimensions of the column diameters.

Table 4. Anatase crystal column widths from sample G measured in four regions of interest (ROI) from STEM cross-section in Figure 13.

ROI	Anatase Column Width (nm)			
	G1	G2	G3	G4
1	80.7 ± 18	101.7 ± 17	221.6 ± 54	136.0 ± 30
2	92.4 ± 8	126.7 ± 31	332.3 ± 3	215.9 ± 20
3	142.4 ± 26	269.8 ± 59	418.3 ± 21	325.5 ± 49
4	315.5 ± 60	397.2 ± 53	419.7 ± 28	511.4 ± 19

The thickness of the nanoplates for the anatase columns G4 were analyzed in the four ROIs specified in Figure 13. Figure 14 shows the variation of the nanoplate thickness plotted against the distance from substrate. The trend shows that the nanostructure thickness increases with the thickness of the films.

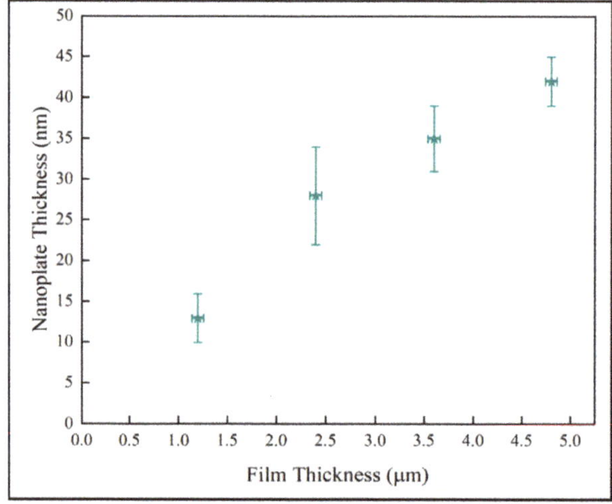

Figure 14. Variation of the mean nanoplate thickness with the distance from substrate in anatase column G4.

4. Discussion

Nanostructured anatase solid coatings have been previously reported by only a few authors [32,33]. Column-like morphologies of TiO_2 are commonly reported for rutile TiO_2 [34] and similar structures have been reported for anatase nanocrystals [35]. Microscale plated-anatase crystals were reported from aerosol CVD [36]. Anatase columns with nanoplates resembling the ones reported in this paper were reported by one other research group who used an atmospheric pressure MOCVD and TTIP + O_2 reactant precursor [37]. These previous studies do not report growth rate or nanostructure dimension, but from descriptions of the processing, we believe that all are relatively rapid growth conditions. High crystal growth rates have been suggested as a factor in generating unstable steps in anatase TiO_2 [38]

which could be an important mechanism in the formation of the nanoplates. The pp-MOCVD flash vaporization produces high mass flux and the temperature of 525 °C is sufficient for rapid reaction of the precursor solution.

Competitive crystal growth in columnar thin films results in increased diameter columns with film thickness and increasing crystallographic alignment [39]. The results in this study are consistent with the current theory. The range of thickness possible with pp-MOCVD allowed us to study the full range of competitive growth. The late stage plateau of crystal column diameter is clearly illustrated in the FIB-STEM sample. The diameter of crystals reaches a certain size where competition stops, but the thickness can continue to increase as crystal diameters remain the same. The orientation of TiO_2 crystals that nucleate on the substrate is random but as the deposition progresses, the fast-growing crystal columns take over and limit the growth of the slower growing crystals. The orientation of the anatase crystals suggests the fast-growing crystals are those that have the (220) planes oriented parallel to the substrate surface [40].

Dynamic scaling theories propose power law interrelationships between microstructural features such as mound spacing, film thickness, interface width (root mean square roughness) and growth time [41]. Figure 15 is the log–log plot of column diameter with film thickness. Scaling exponents from *diameter* $\propto$ *thickness^p* are identified as $p = 0.57$ for shorter deposition times and $p = 0.17$ for longer deposition times. Both linear fits have $R^2 > 0.99$. The reduced exponent for thicker films supports last-stage competitive growth.

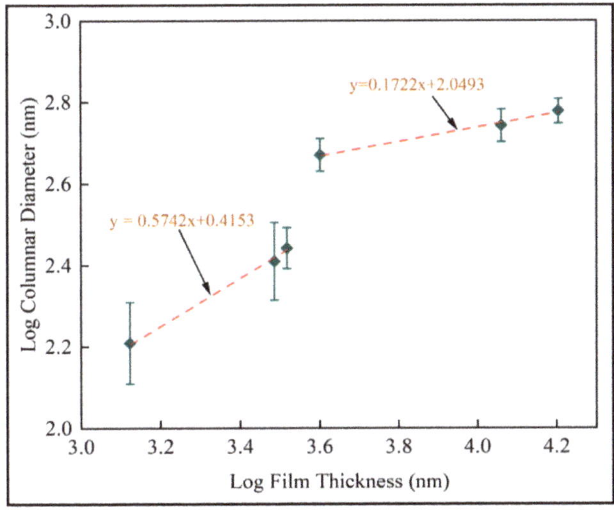

Figure 15. Log transformed plots of column diameter with film thickness for mean measurements on fused silica substrates (samples A–F).

The nanoplate dimension is the most significant result of this study. Our objective is to enhance PCA and antimicrobial activity by inducing nanoscale structure in a solid coating. The results described in this study show that the nano-plate thickness increases with thickness and number of pulses. Figure 16 shows the log transformed plate thickness measurements with film thickness for samples A-F on fused silica. Two scaling regimes are identified from best fit lines, where $R^2 > 0.96$ for both. This shows that plate thickness increases as square root film thickness in the thinnest films investigated here. In thicker films, the plate thickness increases more slowly with film thickness. This is likely to be related to texture and supports the role of competitive growth in film development. We postulate that the increased column diameter provides more space for the nanostructures to grow and hence the resultant increase in their dimensions is observed. The nanoplate dimensions is observed to increase rapidly during the first 600 pulses, but beyond this the nano-dimension begins to plateau. We hypothesize that this is due to the relatively unstable thermodynamic conditions at a lower number of pulses compared to the conditions at a higher number of pulses. It could also be explained by the competitive columnar crystal growth. As the column diameter of the crystals increases, we also observe that each plate has feather-like nanostructures at the edges. This has not been investigated here and high resolution TEM imaging would be required to carry out a similar analysis on the dimensions of these tertiary nanostructures.

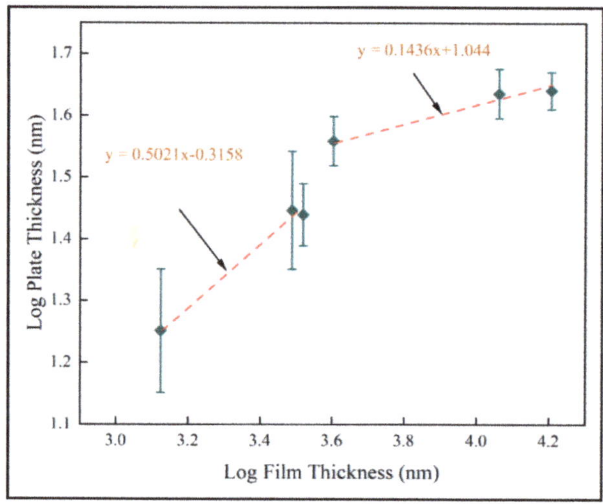

Figure 16. Log transformed plate thickness for mean data on fused silica substrates (samples A–F).

In previous work, we showed that PCA measured by photodegradation of stearic acid had a strong correlation with the surface roughness of mixed phase TiO_2 films grown by pp-MOCVD on fused silica [18]. In that study, the two thickest films, ~1 μm and 4 μm thick, respectively, had the highest surface roughness values and the highest PCA, and correspond to samples A and D in the present work. The smallest nanoplate width, ~18 nm, was measured on sample A. The nanoplates in sample D were more than twice as thick, ~36 nm. This suggests that the PCA is not linearly correlated to the nanoplate width. Other factors could be tertiary branching, texture formation and porosity.

Another phenomenon that has not yet been studied thoroughly is the formation of multi-phase titania columns. Tavares et al. stated that the competitive growth between anatase and rutile phase TiO_2 is dependent on the chemical kinetics that come into play during high-temperature deposition conditions. The authors also report that the quantity of O_2 in the processing environment dictates the phase fractions of anatase and rutile [42]. As the pp-MOCVD processing takes place in a low O_2 environment, the relatively sporadic appearance of rutile phase in these films could be explained. We have previously reported that the rutile columns grow on top of anatase crystals [14]. However, we have also observed that TiO_2 films prepared by pp-MOCVD are composed of both anatase and rutile phases even in the early deposition stages [18]. The next stage is to study the crystallographic orientation and determine the growth conditions that favor the appearance of rutile in the films.

5. Conclusions

This work adds to a body of work aimed at the practical application of nano-engineered TiO_2 solid coatings for antimicrobial touch-surfaces. The pp-MOCVD process has recently provided a processing route to produce nanostructured columnar anatase/rutile/carbon composite TiO_2 coatings. An interesting *mille-feuille* plated nanostructure in anatase columns is investigated as one of the reasons for the observed enhancement of photocatalytic activity. The short migration path from inside the nanoplate to the large specific surface areas is key for achieving an antimicrobial coating. Surface SEM and cross-sectional STEM dimensional measurements have elucidated a relationship between the thickness of the coating and the key nano-dimension. The nano-dimension increases with the column diameter and with the thickness in a similar way as expected with competitive growth mechanisms.

The variation of the nanoplate dimension with sample thickness exhibits two trends. In the early stage depositions ($\leq$600 pulses), the nanoplate dimension increases rapidly with the increase in the number of pulses. The samples deposited for longer durations show a weaker variability of the nanoplate dimensions with sample thickness. The variation of nanoplate dimension appears to plateau with number of pulses over 600 in the same way that column diameter approaches an apparent maximum in the thickest coatings. The relationships reported in this work can provide the basis for engineering of a practical nano-structured TiO_2 coating with the enhanced photocatalytic activity provided by nanoengineering. The coating needs to be thick enough to have durability and long service life, but needs to be thin enough to keep the migration path length low. Because of the relationship described between column diameter in the surface SEM view and the nano-dimension, the quality control of the nano-engineered coating could be much more easily established than if nanoscale observations were required.

Author Contributions: Conceptualization, S.K.; formal analysis R.G., C.B.; writing-original draft preparation, R.G., S.K., C.B.; writing—review and editing, S.K., C.B.; project administration, S.K.; funding acquisition, S.K. All authors have read and agreed to the published version of the manuscript.

Funding: This work was funded by grants from New Zealand Ministry for Business, Innovation and Employment (MBIE) contract UCOX1501.

Acknowledgments: The authors wish to thank Sam Davies Talwar and Darry Lee for project management and sample preparation. We also thank Johann Land for the preparation of part of the sample set. The STEM work was carried out at IMRI, University of California, Irvine. We thank Mecartney, Zheng and the IMRI facility for providing access to the advanced characterization equipment.

Conflicts of Interest: The authors declare no conflict of interest.

References

1. Fujishima, A.; Honda, K. Electrochemical photolysis of water at a semiconductor electrode. *Nature* **1972**, *238*, 37–38. [CrossRef] [PubMed]
2. Morris, D. Pilkington teaches about active self-cleaning glass. *Glass Digest* **2001**, *80*, 36–37.
3. Augustynski, J. The role of the surface intermediates in the photoelectrochemical behavior of anatase and rutile TiO_2. *Electrochim. Acta* **1993**, *38*, 43–46. [CrossRef]
4. Sclafani, A.; Herrmann, J.M. Comparison of the photoelectronic and photocatalytic activities of various anatase and rutile forms of titania in pure liquid organic phases and in aqueous solutions. *J. Phys. Chem.* **1996**, *100*, 13655–13661. [CrossRef]
5. Song, X.M.; Wu, J.M.; Tang, M.Z.; Qi, B.; Yan, M. Enhanced Photoelectrochemical Response of a Composite Titania Thin Film with Single-Crystalline Rutile Nanorods Embedded in Anatase Aggregates. *J. Phys. Chem. C* **2008**, *112*, 19484–19492. [CrossRef]
6. Zhang, J.A.; Huang, Z.H.; Xu, Y.; Kang, F.Y. Carbon-coated TiO_2 composites for the photocatalytic degradation of low concentration benzene. *New Carbon Mater.* **2011**, *26*, 63–69. [CrossRef]
7. Leary, R.; Westwood, A. Carbonaceous nanomaterials for the enhancement of TiO_2 photocatalysis. *Carbon* **2011**, *49*, 741–772. [CrossRef]
8. Linic, S.; Christopher, P.; Ingram, D.B. Plasmonic-metal nanostructures for efficient conversion of solar to chemical energy. *Nat. Mater.* **2011**, *10*, 911–921. [CrossRef]
9. Liu, X.Q.; Iocozzia, J.; Wang, Y.; Cui, X.; Chen, Y.H.; Zhao, S.Q.; Li, Z.; Lin, Z.Q. Noble metal-metal oxide nanohybrids with tailored nanostructures for efficient solar energy conversion, photocatalysis and environmental remediation. *Energy Environ. Sci.* **2017**, *10*, 402–434. [CrossRef]
10. Chen, X.; Mao, S.S. Titanium dioxide nanomaterials: Synthesis, properties, modifications, and applications. *Chem. Rev.* **2007**, *107*, 2891–2959. [CrossRef]

11. Hashimoto, K.; Irie, H.; Fujishima, A. TiO$_2$ photocatalysis: A historical overview and future prospects. *Jpn. J. Appl. Phys. Part 1 Regul. Pap. Brief Commun. Rev. Pap.* **2005**, *44*, 8269–8285. [CrossRef]

12. Schoonman, J. Nanostructured materials in solid state ionics. *Solid State Ion.* **2000**, *135*, 5–19. [CrossRef]

13. Cavallo, C.; Di Pascasio, F.; Latini, A.; Bonomo, M.; Dini, D. Nanostructured Semiconductor Materials for Dye-Sensitized Solar Cells. *J. Nanomater.* **2017**. [CrossRef]

14. Krumdieck, S.P.; Boichot, R.; Gorthy, R.; Land, J.G.; Lay, S.; Gardecka, A.J.; Polson, M.I.J.; Wasa, A.; Aitken, J.E.; Heinemann, J.A.; et al. Nanostructured TiO$_2$ anatase-rutile-carbon solid coating with visible light antimicrobial activity. *Sci. Rep.* **2019**, *9*, 11. [CrossRef] [PubMed]

15. Yang, X.F.; Zhuang, J.L.; Li, X.Y.; Chen, D.H.; Ouyang, G.F.; Mao, Z.Q.; Han, Y.X.; He, Z.H.; Liang, C.L.; Wu, M.M.; et al. Hierarchically Nanostructured Rutile Arrays. Acid Vapor Oxidation Growth and Tunable Morphologies. *ACS Nano* **2009**, *3*, 1212–1218. [CrossRef] [PubMed]

16. Krumdieck, S.; Gorthy, R.; Land, J.G.; Gardecka, A.J.; Polson, M.I.J.; Boichot, R.; Bishop, C.M.; Kennedy, J.V. Titania Solid Thin Films Deposited by pp-MOCVD Exhibiting Visible Light Photocatalytic Activity. *Phys. Status Solidi A-Appl. Mat.* **2018**, *215*, 7. [CrossRef]

17. Mills, A.; Lepre, A.; Elliott, N.; Bhopal, S.; Parkin, I.P.; O'Neill, S.A. Characterisation of the photocatalyst Pilkington Activ™: A reference film photocatalyst? *J. Photochem. Photobiol. A Chem.* **2003**, *160*, 213–224. [CrossRef]

18. Gardecka, A.J.; Polson, M.I.J.; Krumdieck, S.P.; Huang, Y.; Bishop, C.M. Growth stages of nano-structured mixed-phase titania thin films and effect on photocatalytic activity. *Thin Solid Films* **2019**, *685*, 136–144. [CrossRef]

19. Krumdieck, S.; Raj, R. Conversion efficiency of alkoxide precursor to oxide films grown by an ultrasonic-assisted, pulsed liquid injection, metalorganic chemical vapor deposition (pulsed-CVD) process. *J. Am. Ceram. Soc.* **1999**, *82*, 1605–1607. [CrossRef]

20. Krumdieck, S.; Gorthy, R.; Gardecka, A.J.; Lee, D.; Miya, S.S.; Talwar, S.D.; Polson, M.I.J.; Bishop, C. Characterization of photocatalytic, wetting and optical properties of TiO$_2$ thin films and demonstration of uniform coating on a 3-D surface in the mass transport controlled regime. *Surf. Coat. Technol.* **2017**, *326*, 402–410. [CrossRef]

21. Cave, H.M.; Krumdieck, S.P.; Jermy, M.C. Development of a model for high precursor conversion efficiency pulsed-pressure chemical vapor deposition (PP-CVD) processing. *Chem. Eng. J.* **2008**, *135*, 120–128. [CrossRef]

22. ASTM, E. *Standard Test Methods for Determining Average Grain Size*; ASTM International: West Conshohocken, PA, USA, 2004.

23. Downs, R.T.; Hall-Wallace, M. The American Mineralogist crystal structure database. *Am. Miner.* **2003**, *88*, 247–250.

24. Mitchell, D.; Schaffer, B. Scripting-customised microscopy tools for Digital Micrograph™. *Ultramicroscopy* **2005**, *103*, 319–332. [PubMed]

25. Baur, W. Uber die Verfeinerung der Kristallstrukturbestimmung einiger Vertreter des Rutiltyps: TiO$_2$, SnO$_2$, GeO$_2$ und MgF$_2$. *Acta Crystallogr.* **1956**, *9*, 515–520. [CrossRef]

26. Howard, C.J.; Sabine, T.M.; Dickson, F. Structural and thermal parameters for rutile and anatase. *Acta Crystallogr. Sect. B* **1991**, *47*, 462–468. [CrossRef]

27. Gardecka, A.J.; Bishop, C.; Lee, D.; Corby, S.; Parkin, I.P.; Kafizas, A.; Krumdieck, S. High efficiency water splitting photoanodes composed of nano-structured anatase-rutile TiO$_2$ heterojunctions by pulsed-pressure MOCVD. *Appl. Catal. B-Environ.* **2018**, *224*, 904–911. [CrossRef]

28. Ferrari, A.C.; Robertson, J. Raman spectroscopy of amorphous, nanostructured, diamond-like carbon, and nanodiamond. *Philos. Trans. R. Soc. A-Math. Phys. Eng. Sci.* **2004**, *362*, 2477–2512. [CrossRef]

29. Prawer, S.; Nemanich, R.J. Raman spectroscopy of diamond and doped diamond. *Philos. Trans. R. Soc. Lond.* **2004**, *362*, 2537–2565.

30. Liu, Y.; Xue, X.; Fang, H.; Tan, Y.; Chen, R.; Su, Y.; Guo, J. The growth behavior of columnar grains in a TiAl alloy during directional induction heat treatments. *Crystengcomm* **2020**, *22*, 1188–1196. [CrossRef]

31. Thompson, C.V.; Carel, R. Texture development in polycrystalline thin films. *Mater. Sci. Eng. B* **1995**, *32*, 211–219. [CrossRef]

32. Ge, M.Z.; Cao, C.Y.; Huang, J.Y.; Li, S.H.; Chen, Z.; Zhang, K.Q.; Al-Deyab, S.S.; Lai, Y.K. A review of one-dimensional TiO_2 nanostructured materials for environmental and energy applications. *J. Mater. Chem. A* **2016**, *4*, 6772–6801. [CrossRef]

33. Jeevanandam, J.; Barhoum, A.; Chan, Y.S.; Dufresne, A.; Danquah, M.K. Review on nanoparticles and nanostructured materials: History, sources, toxicity and regulations. *Beilstein J. Nanotechnol.* **2018**, *9*, 1050–1074. [CrossRef] [PubMed]

34. Raj, R.; Krumdieck, S.P. A Langmuir-Kinetic Model for CVD Growth from Chemical Precursors. *Chem. Vapor Depos.* **2013**, *19*, 260–266. [CrossRef]

35. An, W.J.; Jiang, D.D.; Matthews, J.R.; Borrelli, N.F.; Biswas, P. Thermal conduction effects impacting morphology during synthesis of columnar nanostructured TiO_2 thin films. *J. Mater. Chem.* **2011**, *21*, 7913–7921. [CrossRef]

36. Wang, W.N.; An, W.J.; Ramalingam, B.; Mukherjee, S.; Niedzwiedzki, D.M.; Gangopadhyay, S.; Biswas, P. Size and Structure Matter: Enhanced CO_2 Photoreduction Efficiency by Size-Resolved Ultrafine Pt Nanoparticles on TiO2 Single Crystals. *J. Am. Chem. Soc.* **2012**, *134*, 11276–11281. [CrossRef]

37. Chen, C.A.; Chen, Y.M.; Huang, Y.S.; Tsai, D.S.; Tiong, K.K.; Liao, P.C. Synthesis and characterization of well-aligned anatase TiO_2 nanocrystals on fused silica via metal-organic vapor deposition. *Crystengcomm* **2009**, *11*, 2313–2318. [CrossRef]

38. Gong, X.Q.; Selloni, A.; Batzill, M.; Diebold, U. Steps on anatase TiO2(101). *Nat. Mater.* **2006**, *5*, 665–670. [CrossRef]

39. He, W.; Mauer, G.; Sohn, Y.J.; Schwedt, A.; Guillon, O.; Vaßen, R. Investigation on growth mechanisms of columnar structured YSZ coatings in Plasma Spray-Physical Vapor Deposition (PS-PVD). *J. Eur. Ceram. Soc.* **2019**, *39*, 3129–3138. [CrossRef]

40. Hawkeye, M.M.; Brett, M.J. Glancing angle deposition: Fabrication, properties, and applications of micro- and nanostructured thin films. *J. Vac. Sci. Technol. A* **2007**, *25*, 1317–1335. [CrossRef]

41. Karabacak, T.; Singh, J.P.; Zhao, Y.P.; Wang, G.C.; Lu, T.M. Scaling during shadowing growth of isolated nanocolumns. *Phys. Rev. B* **2003**, *68*, 125408. [CrossRef]

42. Tavares, C.J.; Vieira, J.; Rebouta, L.; Hungerford, G.; Coutinho, P.; Teixeira, V.; Carneiro, J.O.; Fernandes, A.J. Reactive sputtering deposition of photocatalytic TiO_2 thin films on glass substrates. *Mater. Sci. Eng. B* **2007**, *138*, 139–143. [CrossRef]

Article

Residual Gas Adsorption and Desorption in the Field Emission of Titanium-Coated Carbon Nanotubes

Huzhong Zhang [1,2], Detian Li [1,*], Peter Wurz [2], Yongjun Cheng [1], Yongjun Wang [1], Chengxiang Wang [1], Jian Sun [1], Gang Li [1] and Rico Georgio Fausch [2]

[1] Science and Technology on Vacuum Technology and Physics Laboratory, Lanzhou Institute of Physics, Lanzhou 730000, China; janehuge@126.com (H.Z.); chyj750418@163.com (Y.C.); wyjlxlz@163.com (Y.W.); wcx20130332@163.com (C.W.); zhz252@126.com (J.S.); ligangcasc510@163.com (G.L.)

[2] Physics Institute, University of Bern, 3012 Bern, Switzerland; peter.wurz@space.unibe.ch (P.W.); rico.fausch@space.unibe.ch (R.G.F.)

* Correspondence: lidetian@spacechina.com

Received: 29 July 2019; Accepted: 2 September 2019; Published: 11 September 2019

Abstract: Titanium (Ti)-coated multiwall carbon nanotubes (CNTs) emitters based on the magnetron sputtering process are demonstrated, and the influences of modification to CNTs on the residual gas adsorption, gas desorption, and their field emission characteristic are discussed. Experimental results show that Ti nanoparticles are easily adsorbed on the surface of CNTs due to the "defects" produced by Ar^+ irradiation pretreatment. X-ray photoelectron spectroscopy (XPS) characterization showed that Ti nanoparticles contribute to the adsorption of ambient molecules by changing the chemical bonding between C, Ti, and O. Field emission of CNTs coated with Ti nanoparticles agree well with the Fowler–Nordheim theory. The deviation of emission current under constant voltage is 6.3% and 8.6% for Ti-CNTs and pristine CNTs, respectively. The mass spectrometry analysis illustrated that Ti-coated CNTs have a better adsorption capacity at room temperature, as well as a lower outgassing effect than pristine CNTs after degassing in the process of field emission.

Keywords: carbon nanotubes; residual gas adsorption; residual gas desorption; field emission

---

1. Introduction

Carbon nanotubes (CNTs) have attracted great interest based on their potential industrial applications since their discovery by Iijima [1]. Among a number of proposed applications [2–6], CNTs exhibit important applications in field emission (FE) devices used in vacuum physics, e.g., for ion sources and vacuum gauges. Such CNT field emission devices are most promising industrially, and their practical use is an imminent prospect [7–13]. FE characteristics of nanotubes from a single multi-walled nanotube (MWNT) and MWNT film were first reported by Rinzler et al. and de Heer et al., respectively [14,15]. Subsequently, many CNTs FE emitters were reported to be applied in power microwave sources, ion source, electric propulsion, X-ray tubes, vacuum gauges, and more [16–19].

Generally, CNTs are very sensitive to different types of molecules in their gaseous environment, which is a promising characteristic for sensors. For instance, Pt and Au coatings on the CNTs improved the sensitivity of chemiresistors to indicate the abundance variations of H_2, NH_3, NO_2, etc. [20,21]. However, in virtue of the special gas-sensitivity of CNTs, the adsorption and desorption of residual gas from CNTs could disturb the background vacuum of FE devices, change the work function of CNTs, and affect FE stability and repeatability [22–26]. Due to the change in residual gas coverage on the surface of field emitters, the change in work function plays an important role in electron emission, noise spectrum, and stability [27]. Sergeev et al. also formulated an adsorption–desorption model based on the Kolmogorov equation for field emitters [28]. Dean et al. heated a single-walled nanotube (SWNT) to high temperature to clean the surface, which led to a lower emission current but a better low-frequency stability [23]. Cho et al. had reported that CNTs modified by metallic nanoparticles contributed to extending the FE lifetime of CNTs [29]. Thus, keeping the nanotubes free from residual gases should be an effective way to improve the FE stability and reduce the outgassing of CNTs emitters in ultrahigh vacuum (UHV) electric devices.

In this work, we report on multi-walled carbon nanotube (MWNT) films for possible application in UHV ionization gauges (IG) or mass spectrometers (MS). These CNT films are grown in a thermal chemical vapor deposition (CVD) system onto stainless-steel substrates that were pretreated by the oxidation–reduction method. The as-grown CNTs were then coated by Ti nanoparticles through the magnetron sputtering process. The interaction of CNTs with the residual gas in UHV and FE characteristics of the emitters are experimentally studied and discussed.

2. Materials and Methods

The MWNTs applied in the experiment were prepared on the stainless-steel substrate via an oxidation–reduction treatment [30]. As this is different from existing chemical methods [31], we then used the magnetron sputtering technique for in situ coating of the as-grown pristine CNTs with the Ti nanoparticles. Ar^+ plasma with 1.38 keV energy was applied to bombard the walls of MWNTs to introduce some "defects" on the surface that facilitate the adherence of Ti nanoparticles. In the experiments, the samples were divided into two groups, hereafter referred to as "Ti-CNTs #1" and "Ti-CNTs #2," which were pretreated by Ar^+ plasma for 40 s and 60 s, respectively. Two groups of pretreated MWNTs were then coated inside a vacuum chamber on a sample plate placed 30 cm from a rotating Ti target (4 rpm). The operation parameters of the apparatus were set as follows: microwave power of 200 W, deposition duration of 42 s, argon atmosphere of 0.3 Pa.

Ti-CNTs #1, Ti-CNTs #2 and original pristine CNTs were all characterized by a FEI Tecnai G2 F20 high-resolution transmission electron microscope (HRTEM) (FEI Technologies Inc., Hillsboro, OR, USA) and Peikin-Elmer PHI-5702 multi-functional X-ray photoelectron spectroscopy (XPS) (Physical Electronics Inc., Chanhassen, MN, USA). A diode configuration was established using the CNT films as a cathode against a stainless-steel anode, to evaluate the FE performance of samples. An Inficon 422 quadrupole mass spectrometer (INFICON GmbH, Köln, Germany) was applied to measure the residual gas variation in the UHV chamber.

3. Results and Discussion

3.1. Characteristics of the CNT Samples

TEM micrographs of pristine CNTs without plasma treatment and Ti coating are presented in Figure 1a,b. CNTs with an average diameter around 40–60 nm having hollow and tubular structures are observed in Figure 1a. Most CNTs appeared to be curved or twisted together, and few amorphous carbon or catalyst nanoparticles are observed on the surface of CNTs. The pristine CNTs have a large and wide diameter distribution and good crystallinity, which can be inferred from the Raman spectrum (Figure 1g). The intensity ratio of the G peak at 1568.8 cm^{-1} and D peaks at 1338.7 cm^{-1}, I_G/I_D, for the pristine CNT, is about 1.7, which is comparable to reported high-quality CNTs on stainless-steel alloy substrates [32,33]. The crystalline graphite shell of the pristine CNT consisted of visible lattice fringes parallel to each other (Figure 1b). Figure 1c shows HRTEM images of the Ar^+ plasma-treated MWNTs, which should cause "defects" with more amorphous carbon and tube expansion [34]. The specific defective structures are shown in Figure 1e,f. Except for the adsorption on the normal CNTs outer wall, as shown in Figure 1d, the Ti nanoparticles with a size of 4–6 nm tended to be deposited on sites close to the "defects."

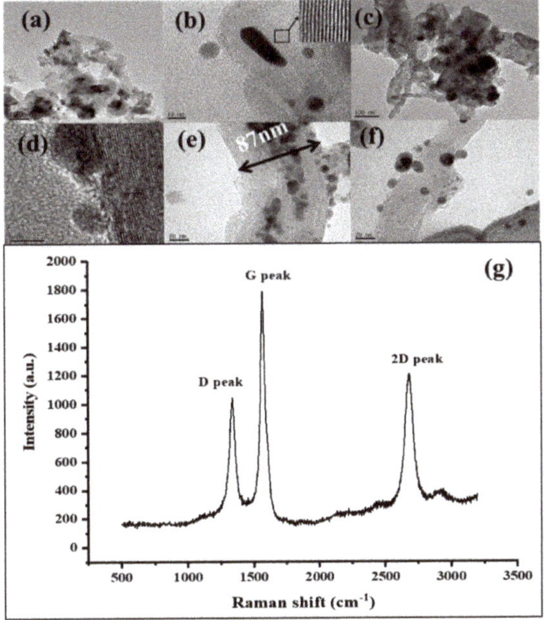

Figure 1. High-resolution transmission electron microscopy (HRTEM) images of (**a**) pristine MWNT, (**b**) magnification of single tube of pristine CNTs (inset: magnification of HRTEM image of outer shells.), (**c**) Ar$^+$-irradiated MWNTs, (**d**) Ti nanoparticles on the normal outer wall of CNTs, (**e**) Ti nanoparticles on the larger-diameter defective CNTs, (**f**) Ti nanoparticles on the "junction" of defective CNTs, (**g**) Raman spectrum of pristine CNTs.

The aggregates of nanoclusters dispersed nonuniformly onto the surface of CNTs, and depend mainly on the Ti nanoparticles' cohesion energy, Ti-CNTs' interface energy, and the diffusion barrier value [35,36]. It is believed that "defects" produced by plasma irradiation create some vacancies that serve as nucleation sites for the adsorption of foreign molecules [37]. The distribution of Ti nanoparticles could be noticed clearly in the two Ti-CNTs samples shown in Figure 2. The Ti nanoparticles are attached onto the surface of tube sidewalls and on the tops randomly, as can be seen in Figure 2a. By extending the Ar$^+$ sputter time to 60 s for Ti-CNT #2 samples, the coverage of Ti nanoparticles increased, and the Ti particles distributed randomly onto the nanotubes as shown in Figure 2b.

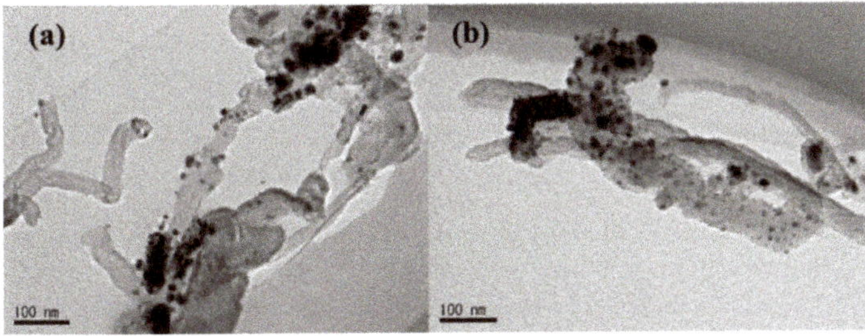

Figure 2. HRTEM of titanium-coated CNTs under the process of magnetron sputtering. (**a**) Ti–CNTs #1 with 40 s irradiation and 42 s DC sputtering. (**b**) Ti-CNTs #2 with 60 s irradiation and 42 s DC sputtering.

The chemical bonds states of CNTs samples have been analyzed. The XPS survey scans of pristine CNTs, Ti-CNTs#1, and Ti-CNTs #2 are shown in Figure 3a–c, which clearly indicate that Ti was successfully deposited on the CNT surface. Carbon bonds were also changed in the processes of Ar^+ irradiation and Ti coating. Figure 3d–f shows the XPS spectra of C 1s states for three samples, in which several deconvoluted peaks, including C–C sp2, C–C sp3, C–O, and C=O bonds, were assigned to the binding energy of 284.4 eV, 285.1 eV, 286.5 eV, and 288 eV, respectively. For the pristine CNTs, the C1s spectrum reveals a high concentration of C–C sp2 hybridized bonds because of the highly crystalline qualities of CNTs. The broad asymmetric tail towards higher binding energy is consistent with C–C sp3 hybridization and C–O bonds. Additionally, the results of Figure 3e,f show that the C–C sp2 hybridized bonds decreased, while C–C sp3 hybridized bonds increased after the Ti deposition. It can be deduced that Ar^+ ions irradiation produced the amorphous "defects" shown in Figure 1. Ti-CNTs #2 undergoing 60 s Ar^+ irradiation with higher C–C hybridized sp3 intensity have more defects on their structure.

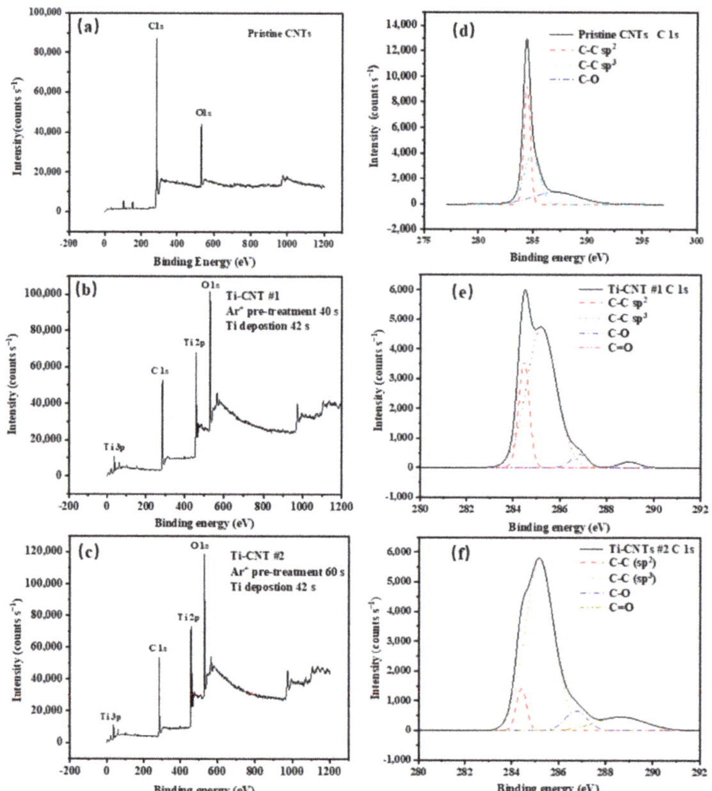

Figure 3. The XPS spectra of different CNT samples: (**a**) pristine CNTs, (**b**) Ti-CNTs #1, (**c**) Ti-CNTs #2, (**d**) C 1s states of Pristine CNTs, (**e**) C 1s states of Ti-CNTs #1, (**f**) C 1s states of Ti-CNTs #2.

The structural "defects" of CNTs provided the adsorption sites for Ti nanoparticles that modify the chemical bonding structure. Because of the oxidation tendency of Ti nanoparticles, the chemical binding among Ti, C, and O changed. The alteration of O 1s states is shown in Figure 4. For the pristine CNTs, O–O at a binding energy of 532 eV (originating from an adsorbed O_2 molecule) is observed. Moreover, some O atoms formed C–O bonds and C=O bonds at a binding energy of 531.5 eV and 532.5 eV, respectively, whereas, for Ti-CNTs #1 and Ti-CNTs #2, particular chemical shifts of the main peak are noted. This could be caused by Ti–O binding from the titanium oxide at a binding energy of

530.2 eV. Meanwhile, Ti–O binding could prevent the formation of C–O bonds and C=O bonds, which are the key factors that affect the FE stability and repeatability of CNTs [29].

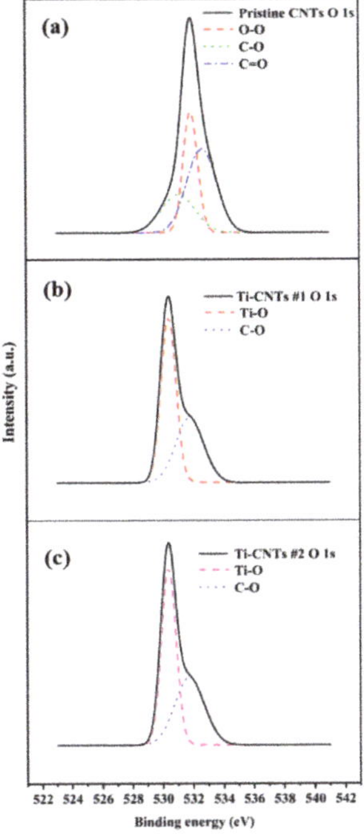

Figure 4. The XPS spectra of O 1s states of (**a**) pristine CNTs, (**b**) Ti-CNT #1, and (**c**) Ti-CNT #2.

Figure 5 shows the Ti 2p states of Ti-CNT #1 and Ti-CNT#2. Both Ti-CNTs had the peaks of Ti bonds corresponding to Ti 2p3/2 and Ti 2p1/2 spin-orbital doublets at 458.5 eV and 464.8 eV, respectively. The sharp and symmetrical peaks of Ti 2p states were close to those of bulk TiO$_2$, indicating the presence of Ti clusters [38]. Combining these with the Ti–O bonds at the binding energy of 530.2 eV in Figure 4, it can be deduced that Ti was significantly oxidized. The Ti-CNTs' surface layers, composed of Ti–O bonds, should protect the CNTs' structure from being damaged by molecules of the ambient gas, i.e., O$_2$ and H$_2$O, which could be confirmed from the chemical shifts in Figure 4.

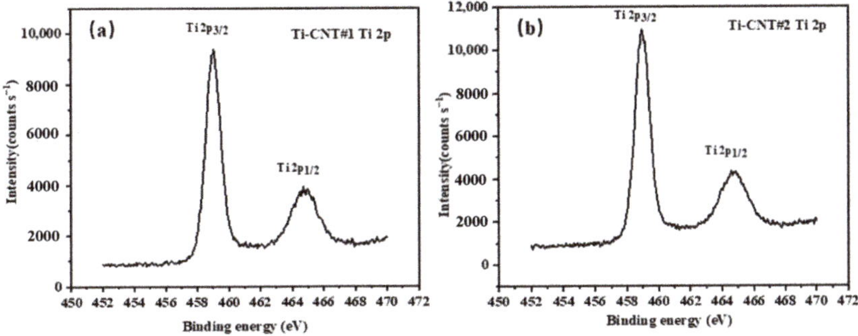

Figure 5. The XPS spectrum of Ti 1p states of (**a**) Ti-CNT #1 and (**b**) Ti-CNT #2.

Using diode devices, pristine CNTs, Ti-CNT #1, and Ti-CNT #2 were fabricated and tested in UHV. The CNTs samples of 28.3 mm^2 emission area were grounded with a ballast resistor of 24 kΩ connected in series. It was kept 150 μm away from the anode. To ensure reliable FE operations, all the samples were put in a vacuum at ~10^{-6} Pa to be conditioned in advance at a FE current of ~30 μA for 1 h. Figure 6 shows the FE current density–electric field (*J–E*) curves and the corresponding Fowler–Nordheim (F–N) plots. The FE of pristine CNTs shows a lower turn-on field (2.6 V/μm) and higher emission current than that of Ti-coated CNTs (5.0 V/μm for Ti-CNT #1 and 6.9 V/μm for Ti-CNT #2) in Figure 6a. The higher turn-on field could be explained by the loss of chemisorbed surface states and accompanying resonant-enhanced tunneling current due to the adsorption of residual gas by Ti nanoparticles rather than CNTs [22]. Moreover, the presence of H dipoles is well known to reduce the work function of CNTs, but the high adsorption capacity for H of Ti decreases the number of H dipoles on the surface of the CNTs [39,40]. Meanwhile, due to partial oxidization, the Ti–O bond was recognized to increase the work function of Ti-coated CNTs. It was theoretically estimated to increase by ~0.6 eV [29]. In addition, plasma treatment may also result in a radius increase of CNT tips, which weakens the field enhancement effect. The tube expansion shown in Figure 1c, in agreement with earlier work [34], could result in a smaller field enhancement factor. Both the increase of work function and the decrease of field enhancement factor lead to the slope change of fitted F–N curves in Figure 6b.

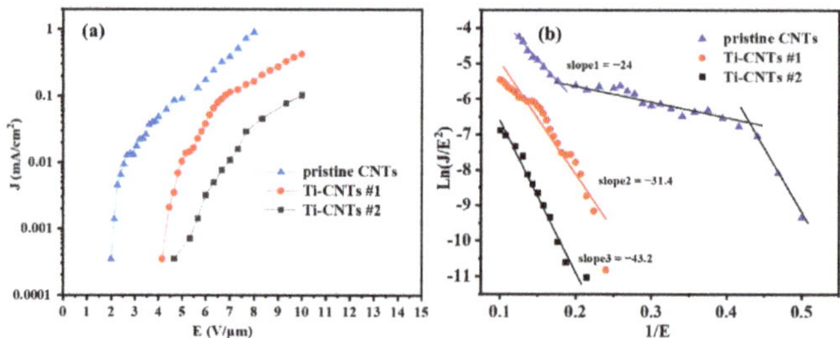

Figure 6. Field emission performance of different emitters. (**a**) J–E electron emission properties, and (**b**) F–N plots of these data.

In Figure 6b, the pristine CNTs indicate typical FE characteristics, with three emission stages of FE, including adsorption-dominated emission at low current range, intermediate range, and intrinsic emission at high current range [23,41,42]. Ti-coated CNTs have better agreement with the F–N theory than pristine CNTs, as shown by the "orthodoxy test" proposed by Forbes [43,44]. Assuming that all the samples have intrinsic emission, for estimated work function 4.8 eV and calculated slopes (Figure 6b), the scaled barrier field *f* ranges are (0.15–0.62), (0.25–0.59), and (0.20–0.43) for pristine CNTs, Ti-CNTs #1, and Ti-CNTs #2, respectively. For the orthodox emission, the values *f* are supposed to range in 0.15–0.44. Thus, we conclude that the discrete Ti compound layer adsorbed most of the residual gas and kept the surface of nanotubes clean, which contributed to such orthodox FE without adsorption dominated emission for Ti-CNTs #2. However, the adsorbates modified the work function to cause "unorthodox" emission for Ti-CNT #1 and pristine CNTs.

To make a more definitive comparison of three CNT samples, the stability tests were conducted in the 10^{-7} Pa UHV chamber and the emission current was recorded under constant voltage mode. As shown in Figure 7a, a significant drop in current occurred in the first 15 min of conditioning, resulting from the direct interaction of residual gas with CNTs and their structural degradation. Along with the desorption of residual gas, the emission current gradually became stable. This did not happen to the Ti-coated CNTs in 4 h of stability tests. As shown in Figure 7b, even after the initial conditioning, there are still discontinuous jumps in the emission current for pristine CNTs, which could lead to the electric disturbance or failure of devices. Neglecting the significant discontinuous jumps, the average emission current of pristine CNTs is 150.6 μA (anode voltage: 1644 V) with a relative standard deviation (RSD) of 8.6%. In contrast, Ti-coated CNTs showed better stability. The emission current is kept to 127.5 µA (anode voltage 1830 V) and 99.2 µA (anode voltage: 2500 V), with a smaller RSD of 7.2% and 6.3% for Ti-CNT #1 and Ti-CNT #2, respectively. More importantly, it could be seen that the emission current of Ti-coated CNTs did not show any abrupt jumps. The emission current increases slightly and becomes stable gradually. The stability improvement of Ti-coated CNTs is mainly caused by the desorption of residual gas, as well as the Ar^+ plasma treatment [27,45]. Pristine CNTs with adsorbed residual gas influence chemisorbed surface states and increase the 1/f noise component. Employing the flicker noise interpretation described by Schottky [46], the 1/f noise spectrum of the gaseous adsorbate can be described by the adsorption–desorption model [47]. The flicker noise was proved to be dependent on the residual gas coverage, where the 1/f part is enlarged owing to the shot noise [27].

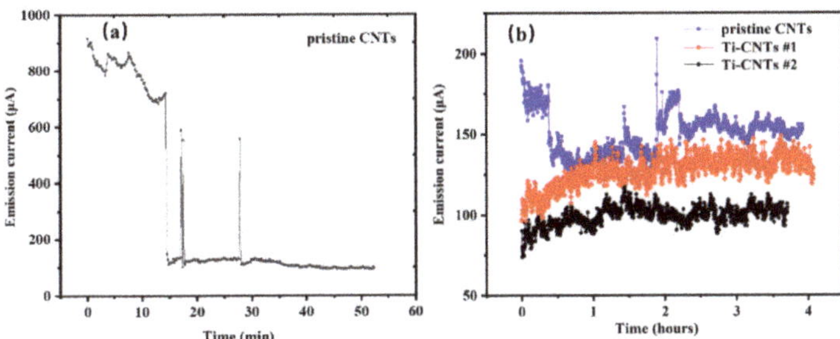

Figure 7. Emission current summary of Ti-coated CNTs and pristine CNTs over time. (**a**) Aging conditioning of pristine CNTs during a short period. (**b**) Stability comparison.

3.2. Gas Adsorption of Different CNTs at Room Temperature

To further analyze the influence of Ti-coated CNTs on the ultrahigh vacuum (UHV), pristine CNTs, Ti-CNTs #1, and Ti-CNTs #2 were successively installed in the UHV chamber via feedthrough. The composition of background residual gases was first measured while the vacuum chamber was degassed at 260 °C for 24 h, as well as the IG (extractor gauge, Leybold Corp., Cologne, Germany) degassed several times. Figure 8 shows the typical temporal development of residual gas species in the UHV chamber of 3×10^{-6} Pa. It is also well known that CO, CO_2, and CH_4 are the typical residual gas components or chemical reaction products desorbed from the IG and vacuum chamber walls, dominated by H_2 and H_2O. While the partial pressure in the chamber is stable, the IG, using an $Ir–Y_2O_3$ filament, was degassed via electron bombardment at an emission current of ~80 mA. As shown in Figure 8, electron-stimulated desorption (ESD) of CO^+ increases dramatically during degassing of the filament, but other gases, including $CO_2{}^+$, $CH_4{}^+$, and $H_2{}^+$, rise only slightly [48].

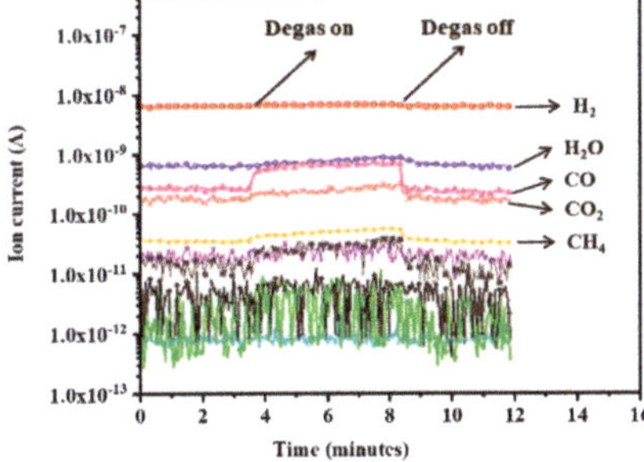

Figure 8. Residual gas composition of UHV chamber background after 24 h degassing, corresponding to a total pressure of 3×10^{-6} Pa.

The pristine CNTs emitter was placed into the vacuum chamber without FE operation at room temperature. Figure 9 shows the gas adsorption capability of pristine CNTs in the vacuum chamber. The vacuum chamber was not degassed at high temperature but pumped over a long period of time. Although the H_2O was the main component, the background partial pressure of the residual gases H_2, CO, and CO_2 decreased apparently. The weak van der Waals interaction between pristine CNTs and residual gases played a key role in the results.

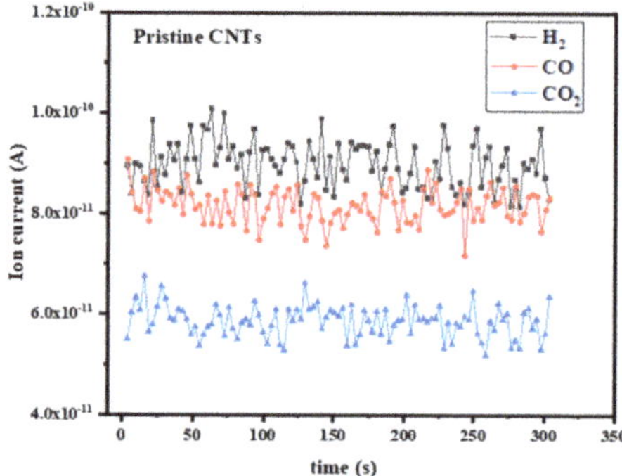

Figure 9. Background residual gas in UHV at room temperature for pristine CNTs without FE operation, corresponding to a total pressure of 1×10^{-6} Pa.

For the Ti-CNTs #1 and Ti-CNTs #2, Ti atoms (a transition metal) were prone to react with H_2 and oxocarbons. Ti nanoparticles slightly changed the atomic level structure of CNTs in combination with residual gases. In the case of no electron emission, the adsorption performance on Ti-CNTs is shown in Figure 10. Both H_2^+ signals decreased to the level of 10^{-11} A at room temperature, while Ti-CNTs #1 and Ti-CNTs #2 were placed in a vacuum chamber. In particular, owing to more Ti coverage, Ti-CNT #2 showed a larger adsorption capacity than Ti-CNTs #1 for residual gases. The CO^+ and CO_2^+ signals were also distinctly reduced for Ti-CNT #2. This means that the residual gas was mainly adsorbed by Ti compound to keep CNTs away from the gas adsorbates. The results could also explain the good intrinsic FE characteristic of Ti-CNT #2.

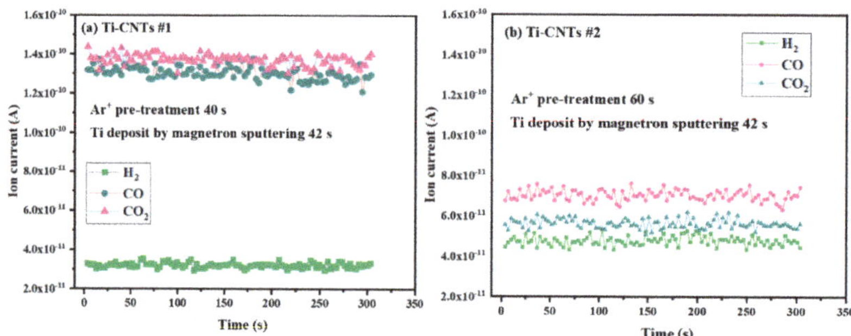

Figure 10. Background residual gas of titanium coated CNTs in UHV of 1×10^{-6} Pa at room temperature and without FE operation. (**a**) Ti-CNTs #1, (**b**) Ti-CNTs #2.

3.3. Gas Desorption of CNTs under Field Emission

The gas desorption was tested in the processes of FE for pristine CNTs, Ti-CNT #1, and Ti-CNT #2. Variations in the main reactive gases in UHV, i.e., H_2, CO, and CO_2, were detected, as shown in Figure 11. In Figure 11a, the H_2^+ signal of pristine CNTs, with a background level of 5.97×10^{-11} A, increases to 1.74×10^{-8} A, along with the emission current rising from 0 µA to 427 µA. It is indicated that the H_2 adsorbed on the surface of pristine CNTs was released slowly along with the increase in the FE current, which may influence the FE performance and UHV. In contrast, H_2^+ in the Ti-CNTs #1 and Ti-CNTs #2 were released rapidly under a low current. The ion signal of H_2^+ increases from 3.22×10^{-11} A to 5.95×10^{-8} A when the emission current varies from 0 μA to 123 μA for Ti-CNTs #1. For Ti-CNTs #2, when the FE current increases from 0 μA to 30 μA, H_2 dissociates faster than the two above-mentioned samples.

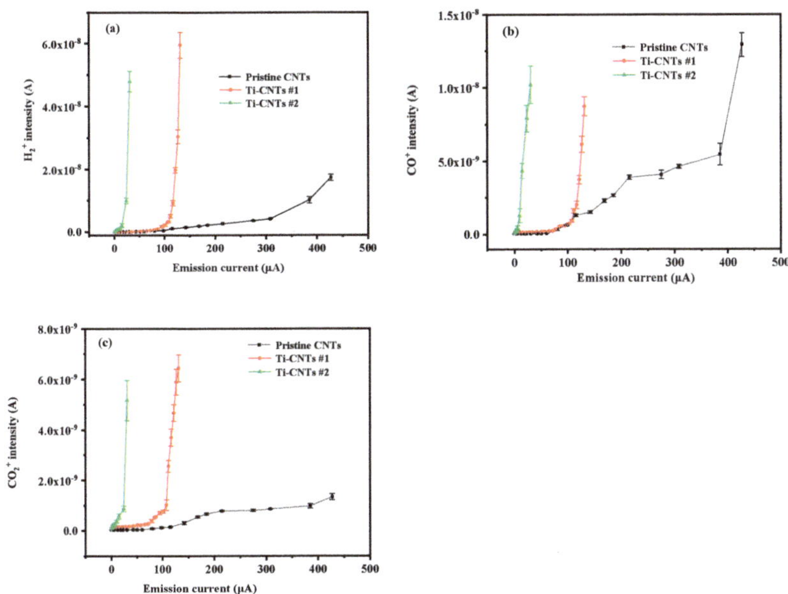

Figure 11. Gas desorption along with variation of FE current for pristine CNTs, CNTs #1 and CNTs #2. (**a**) H_2^+ intensity, (**b**) CO^+ intensity, (**c**) CO_2^+ intensity.

In Figure 11b, CO does not show a significant increase for the pristine CNTs until the emission current is higher than ~60 μA, and therefore most of the released CO should be typical ESD ions from the stainless-steel anode [48]. By contrast, CO instantly increased to the same level as pristine CNTs for Ti-modified samples, originating from ESD and gas dissociation effects. Moreover, as shown in Figure 11c, CO_2 increases slightly for pristine CNTs due to the low background pressure of CO_2 in UHV. However, CO_2 is prone to combine with Ti and Ti oxides, so the CO_2 on Ti-CNT #1 and Ti-CNT#2 was also dissociated from the surface of Ti nanoparticles more than pristine CNTs.

A knowledge of the adsorption–desorption properties of Ti-CNTs would be helpful to reduce the outgassing of CNTs emitters in the UHV electronic devices. The Ti-CNTs #2 and pristine CNTs were placed into a UHV chamber to conduct degassing experiments. The pressure variations, along with the emission current, were recorded in Table 1. When the emission current was set to 500 μA, Joule heating of the emitters caused degassing of the CNTs and made the pressure rise from 6.98×10^{-7} Pa to 7.73×10^{-7} Pa for Ti-CNTs #2 and from 4.16×10^{-7} Pa to 4.51×10^{-7} Pa for pristine CNTs. Then the current was stabilized to 500 μA for 10 min of degassing. The pressure for Ti-coated CNTs decreased from 7.73×10^{-7} Pa to 6.96×10^{-7} Pa; in contrast, for pristine CNTs, it increased from 4.51×10^{-7} Pa to 4.55×10^{-7} Pa. Then the emission current of both emitters was adjusted to a typical current (200 μA) for IG and MS. There was a significant difference between the two emitters 10 min later. The final working pressure variation Δp ($p_2' - p_u$) is 6×10^{-9} Pa and -2×10^{-8} Pa for the pristine CNTs and Ti-CNTs #2, respectively. The results support our assertion that the low outgassing property of Ti-CNTs will limit the background disturbance and allow for accurate measurements in a UHV environment [49].

Table 1. Pressure variation during the degassing of Ti-CNTs #2 and pristine CNTs.

Pressure (Time)	Ti-CNT#2 (Pa)	Pristine CNTs (Pa)	Emission Current (μA)
p_u (t_0)	6.98×10^{-7}	4.16×10^{-7}	0
p_1 (t_1)	7.73×10^{-7}	4.51×10^{-7}	500
p_1' (t_1 + 10 min)	6.96×10^{-7}	4.55×10^{-7}	500
p_2 (t_2)	6.92×10^{-7}	4.27×10^{-7}	200
p_2' (t_2 + 10 min)	6.78×10^{-7}	4.22×10^{-7}	200
Δp ($p_2' - p_u$)	-2.00×10^{-8}	6.00×10^{-9}	200

4. Conclusions

We investigated the influence of Ti coating of MWNTs on the adsorption of residual gas in a UHV environment and its later release during operation. Ti-coated CNTs utilizing magnetron sputtering technology enhanced the adsorption capacity of residual gas based on the chemical activity of Ti nanoparticles. The partial pressure of H_2, CO, and CO_2 was reduced significantly by Ti-CNTs at room temperature. The process could prevent the contamination of the CNTs surface, which would lead to good FE consistency with the F–N theory and excellent FE stability as well. Ti-CNTs prevent the electric noise or failure of devices caused by discontinuous jumps in the current emission. Moreover, degassing by Joule heating in the process of FE allows Ti nanoparticles to release adsorbed gases rapidly, which provides a way to effectively degas the CNTs emitter. It is thus a promising field emitter and should find applications in UHV measurement devices.

Ti coating of CNTs has been shown to have a potential use in practical applications involving CNTs as a FE electron source. The stability of Ti-CNTs emitters over thousands of hours and their outgassing properties in IG and QMS should be studied in further research work.

Author Contributions: H.-Z.Z. carried out the experiment and wrote the original draft of the manuscript. D.-T.L. and H.-Z.Z. conceptualized the article and created the methodology. P.W. contributed to the discussion and revision of manuscript. Y.-J.C. supervised the project. H.-Z.Z., Y.-J.W., C.-X.W. and J.S. took part in the experiments, including preparation of CNTs, Ti nanoparticles coating, characterization of morphology, FE tests and residual gas analyzation. G.L. conducted the stability evaluation of samples. R.G.F. took part in the data analyzation and discussed the results. All authors read and approved the final manuscript.

Funding: This research was funded by the National Natural Science Foundation of China (grants 61601211, 61671226, 61627805, and 61771228) and Swiss National Science Foundation (SNSF, 200020-184657/1).

Acknowledgments: The authors are grateful to Dong Changkun for the helpful discussions, and also thank Zhangbin, Guo Lei, and Zhang Shangwei for technical support.

References

1. Iijima, S. Synthesis of Carbon Nanotubes. *Nature* **1991**, *354*, 56–58. [CrossRef]
2. Baughman, R.H.; Zakhidov, A.A.; de Heer, W.A. Carbon nanotubes—The route toward applications. *Science* **2002**, *297*, 787–792. [CrossRef] [PubMed]
3. Dai, H.; Hafner, J.H.; Rinzler, A.G.; Colbert, D.T.; Smalley, R.E. Nanotubes as nanoprobes in scanning probe microscopy. *Nature* **1996**, *384*, 147–150. [CrossRef]
4. Dillon, A.C.; Jones, K.M.; Bekkedahl, T.A.; Kiang, C.H.; Bethune, D.S.; Haben, M.J. Storage of hydrogen in single-walled carbon nanotubes. *Nature* **1997**, *386*, 377–379. [CrossRef]
5. di Bartolomeo, A.; Yang, Y.; Rinzan, M.B.M.; Boyd, A.K.; Barbara, P. Record Endurance for Single-Walled Carbon Nanotube–Based Memory Cell. *Nanoscale Res. Lett.* **2010**, *5*, 1852–1855. [CrossRef] [PubMed]
6. Giordano, C.; Filatrella, G.; Sarno, M.; Di Bartolomeo, A. Multi-walled carbon nanotube films for the measurement of the alcoholic concentration. *Micro Nano Lett.* **2019**, *14*, 304–308. [CrossRef]
7. Milne, W.I.; Teo, K.B.K.; Amaratunga, G.A.J.; Legagneux, P.; Gangloff, L.; Schnell, J.P.; Semet, V.; Binh, V.T.; Groening, O. Carbon nanotubes as field emission sources. *J. Mater. Chem.* **2004**, *14*, 933–943. [CrossRef]

8. Liang, W.; Bockrath, M.; Bozovic, D.; Hafner, J.H.; Tinkham, M.; Park, H. Fabry—Perot interference in a nanotube electron waveguide. *Nature* **2001**, *411*, 665–669. [CrossRef] [PubMed]

9. Yao, Z.; Kane, C.L.; Dekker, C. High-Field Electrical Transport in Single-Wall Carbon Nanotubes. *Phys. Rev. Lett.* **2000**, *84*, 2941–2944. [CrossRef] [PubMed]

10. Lei, W.; Zhang, X.; Lou, C.; Zhao, Z.; Wang, B. Very high field-emission current from a carbon-nanotube cathode with a pulse driving mode. *IEEE Electron Device Lett.* **2009**, *30*, 571–573. [CrossRef]

11. Zhang, Q.; Wang, X.; Meng, P.; Yue, H.; Zheng, R.; Wu, X.; Cheng, G.-A. High current density and low emission field of carbon nanotube array microbundle. *Appl. Phys. Lett.* **2018**, *112*, 013101. [CrossRef]

12. Li, D.; Wang, Y.; Cheng, Y.; Feng, Y.; Zhao, L.; Zhang, H.; Sun, J.; Dong, C. An overview of ionization gauges with carbon nanotube cathodes. *J. Phys. D Appl. Phys.* **2015**, *48*, 473001. [CrossRef]

13. Wilfert, S.; Edelmann, C. Field emitter-based vacuum sensors. *Vacuum* **2012**, *86*, 556–571. [CrossRef]

14. Rinzler, A.G.; Hafner, J.H.; Nikolaev, P.; Lou, L.; Kim, S.G.; Tomanek, D.; Nordlander, P.; Colbert, D.T.; Smalley, R.E. Unraveling nanotubes: Field emission from an atomic wire. *Science* **1995**, *269*, 1550–1553. [CrossRef]

15. De Heer, W.A.; Chatelain, A.; Ugarte, D. A Carbon Nanotube Field-Emission Electron Source. *Science* **1995**, *270*, 1179–1180. [CrossRef]

16. Lee, K.J.; Hong, N.T.; Lee, S.; You, D.-W.; Jung, K.-W.; Yang, S.S. Simple fabrication of micro time-of-flight mass spectrometer using a carbon nanotube ionizer. *Sens. Actuators B* **2017**, *243*, 394–402. [CrossRef]

17. Teo, K.B.; Minoux, E.; Hudanski, L.; Peauger, F.; Schnell, J.P.; Gangloff, L.; Legagneux, P.; Dieumegard, D.; Amaratunga, G.A.; Milne, W.I. Microwave devices: Carbon nanotubes as cold cathodes. *Nature* **2005**, *437*, 968. [CrossRef]

18. Okawa, Y.; Kitamura, S.; Kawamoto, S.; Iseki, Y.; Hashimoto, K.; Noda, E. An experimental study on carbon nanotube cathodes for electrodynamic tether propulsion. *Acta Astronaut.* **2007**, *61*, 989–994. [CrossRef]

19. Lei, W.; Zhu, Z.; Liu, C.; Zhang, X.; Wang, B.; Nathan, A. High-current field-emission of carbon nanotubes and its application as a fast-imaging X-ray source. *Carbon* **2015**, *94*, 687–693. [CrossRef]

20. Penza, M.; Cassano, G.; Rossi, R.; Alvisi, M.; Rizzo, A.; Signore, M.A. Enhancement of sensitivity in gas chemiresistors based on carbon nanotube surface functionalized with noble metal (Au, Pt) nanoclusters. *Appl. Phys. Lett.* **2007**, *90*, 173123. [CrossRef]

21. Kong, J.; Chapline, M.G.; Dai, H. Functionalized Carbon Nanotubes for Molecular Hydrogen Sensors. *Adv. Mater.* **2001**, *13*, 1384–1386. [CrossRef]

22. Dean, K.A.; Chalamala, B.R. Field emission microscopy of carbon nanotube caps. *J. Appl. Phys.* **1999**, *85*, 3832–3836. [CrossRef]

23. Dean, K.A.; von Allmen, P.; Chalamala, B.R. Three behavioral states observed in field emission from single-walled carbon nanotubes. *J. Vac. Sci. Technol. B* **1999**, *17*, 1959–1969. [CrossRef]

24. Dong, C.; Luo, H.; Cai, J.; Wang, F.; Zhao, Y.; Li, D. Hydrogen sensing characteristics from carbon nanotube field emissions. *Nanoscale* **2016**, *8*, 5599–5604. [CrossRef] [PubMed]

25. Grzebyk, T.; Górecka-Drzazga, A. Miniature ion-sorption vacuum pump with CNT field-emission electron source. *J. Micromech. Microeng.* **2013**, *23*, 015007. [CrossRef]

26. Lotz, M.; Wilfert, S.; Kester, O. Development of a field emitter-based extractor gauge for the operation in cryogenic vacuum environments. In Proceedings of the IPAC2014, Dresden, Germany, 16–20 June 2014; pp. 2320–2322.

27. Knápek, A. *Methods of Preparation and Characterisation of Experimental Field-Emission Cathodes*; Brno University of Technology: Brno, Czech Republic, 2013.

28. Sergeev, E.; Knápek, A.; Grmela, L.; Šikula, J. Noise diagnostic method of experimental cold field-emission cathodes. In Proceedings of the 2013 22nd International Conference IEEE Noise and Fluctuations (ICNF), Montpellier, France, 24–28 June 2013; pp. 1–4.

29. Cho, Y.; Kim, C.; Moon, H.; Choi, Y.; Park, S.; Lee, C.-K.; Han, S. Electronic Structure Tailoring and Selective Adsorption Mechanism of Metal-coated Nanotubes. *Nanoletters* **2008**, *8*, 81–86. [CrossRef] [PubMed]

30. Wang, Y.; Li, D.; Sun, W.; Sun, J.; Li, G.; Zhang, H.; Xi, Z.; Pei, X.; Li, Y.; Cheng, Y. Synthesis and field electron emission properties of multi-walled carbon nanotube films directly grown on catalytic stainless steel substrate. *Vacuum* **2018**, *149*, 195–199. [CrossRef]

31. Ma, X.; Li, X.; Lun, N.; Wen, S. Synthesis of gold nano-catalysts supported on carbon nanotubes by using electroless plating technique. *Mater. Chem. Phys.* **2006**, *97*, 351–356. [CrossRef]

32. Yamagiwa, K.; Ayato, Y.; Kuwano, J. Liquid-phase synthesis of highly aligned carbon nanotubes on preheated stainless steel substrates. *Carbon* **2016**, *98*, 225–231. [CrossRef]

33. Latorre, N.; Cazana, F.; Sebastian, V.; Royo, C.; Romeo, E.; Centeno, M.A.; Monzón, A. Growth of carbonaceous nanomaterials over stainless steel foams effect of activation temperature. *Catal. Today* **2016**, *273*, 41–49. [CrossRef]

34. Kim, H.M.; Kim, H.S.; Park, S.K.; Joo, J.; Lee, T.J.; Lee, C.J. Morphological change of multiwalled carbon nanotubes through high-energy (MeV) ion irradiation. *J. Appl. Phys.* **2005**, *97*, 026103. [CrossRef]

35. Jiang, L.; Liu, P.; Zhang, L.; Liu, C.; Zhang, L.; Fan, S. The adsorption state and the evolution of field emission properties of graphene edges at different temperatures. *RSC Adv.* **2018**, *8*, 31830–31834. [CrossRef]

36. Young, T. An essay on the cohesion of fluids. *Philos. Trans. R. Soc. Lond.* **1805**, *95*, 65–87. [CrossRef]

37. Charlier, J.-C. Defects in Carbon Nanotubes. *Acc. Chem. Res.* **2002**, *35*, 1063–1069. [CrossRef] [PubMed]

38. Bogle, K.A.; Bachhav, M.N.; Deo, M.S.; Valanoor, N.; Ogale, S.B. Enhanced nonvolatile resistive switching in dilutely cobalt doped TiO_2. *Appl. Phys. Lett.* **2009**, *95*, 203502–203503. [CrossRef]

39. Murray, P.T.; Back, T.C.; Cahay, M.M.; Fairchild, S.B.; Maruyama, B.; Lockwood, N.P.; Pasquali, M. Evidence for adsorbate-enhanced field emission from carbon nanotube fibers. *Appl. Phys. Lett.* **2013**, *103*, 053113. [CrossRef]

40. Ago, H.; Kugler, T.; Cacialli, F.; Salaneck, W.R.; Shaffer, M.S.P.; Windle, A.H.; Friend, R.H. Work functions and surface functional groups of multiwall carbon nanotubes. *J. Phys. Chem. B* **1999**, *103*, 8116–8121. [CrossRef]

41. Janas, D.; Koziol, K.K.K. The influence of metal nanoparticles on electrical properties of carbon nanotubes. *Appl. Surf. Sci.* **2016**, *376*, 74–78. [CrossRef]

42. Brodie, I.; Spindt, C.A. Vacuum Microelectronics. *Adv. Electron. Electron Phys.* **1992**, *83*, 1–106.

43. Forbes, R.G. Use of a Spreadsheet to Test for Lack of Field Emission Orthodoxy. In Proceedings of the 2014 Tenth International Vacuum Electron Sources Conference (IVESC), Saint-Petersburg, Russia, 30 June–4 July 2014.

44. Forbes, R.G. Development of a simple quantitative test for lack of field emission orthodoxy. *Proc. R. Soc. A* **2013**, *469*, 20130271. [CrossRef]

45. Evtukh, A.; Hartnagel, H.; Yilmazoglu, O.; Mimura, H.; Pavlidis, D. *Vacuum Nanoelectronic Devices: Novel Electron Sources and Applications*, 1st ed.; Wiley: Hoboken, NJ, USA, 2015; pp. 315–374. ISBN 9781119037989.

46. Schottky, W. Small-Shot Effect and Flicker Effect. *Phys. Rev.* **1926**, *28*, 74. [CrossRef]

47. Kleint, C. Experimente zum Funkelrauschen bei Feldemission und Vergleich mit theoretischen Vorstellungen. *Ann. Phys.* **1963**, *10*, 309–320. [CrossRef]

48. Redhead, P.A. The first 50 years of electron stimulated desorption (1918–1968). *Vacuum* **1997**, *48*, 585–596. [CrossRef]

49. Redhead, P.A. UHV and XHV pressure measurements. *Vacuum* **1993**, *44*, 559–564. [CrossRef]

 *materials*

Article

Use of a New Non-Pyrophoric Liquid Aluminum Precursor for Atomic Layer Deposition

Xueming Xia [1], Alaric Taylor [2], Yifan Zhao [3], Stefan Guldin [2] and Chris Blackman [1],*

[1] Department of Chemistry, University College London, 20 Gordon Street, London WC1H 0AJ, UK;
xueming.xia.16@ucl.ac.uk
[2] Department of Chemical Engineering, University College London, Torrington Place, London WC1E 7JE, UK;
alaric.taylor@ucl.ac.uk (A.T.); s.guldin@ucl.ac.uk (S.G.)
[3] Department of Life Sciences, Imperial College London, South Kensington Campus, London SW7 2AZ, UK;
yifan.zhao15@imperial.ac.uk
* Correspondence: c.blackman@ucl.ac.uk; Tel.: +44-207-679-4703

Received: 2 April 2019; Accepted: 30 April 2019; Published: 2 May 2019

Abstract: An Al_2O_3 thin film has been grown by vapor deposition using different Al precursors. The most commonly used precursor is trimethylaluminum, which is highly reactive and pyrophoric. In the purpose of searching for a more ideal Al source, the non-pyrophoric aluminum tri-sec-butoxide ($[Al(O^sBu)_3]$, ATSB) was introduced as a novel precursor for atomic layer deposition (ALD). After demonstrating the deposition of Al_2O_3 via chemical vapor deposition (CVD) and 'pulsed CVD' routes, the use of ATSB in an atomic layer deposition (ALD)-like process was investigated and optimized to achieve self-limiting growth. The films were characterized using spectral reflectance, ellipsometry and UV-Vis before their composition was studied. The growth rate of Al_2O_3 via the ALD-like process was consistently 0.12 nm/cycle on glass, silicon and quartz substrates under the optimized conditions. Scanning electron microscopy and transmission electron microscopy images of the ALD-deposited Al_2O_3 films deposited on complex nanostructures demonstrated the conformity, uniformity and good thickness control of these films, suggesting a potential of being used as the protection layer in photoelectrochemical water splitting.

Keywords: chemical vapor deposition; atomic layer deposition; aluminum oxide; aluminum tri-sec-butoxide; thin film

1. Introduction

As a dielectric metal oxide, Al_2O_3 exhibits a high transparency, large bandgap and excellent electrical insulation properties, hence it is widely applied in electronic devices and electrochemistry, e.g., as a protection layer [1–6]. Commonly used methods for making Al_2O_3 films include sol-gel, sputtering, evaporation, physical vapor deposition (PVD), chemical vapor deposition (CVD) and atomic layer deposition (ALD) [7–11]. CVD is a widely used thin film deposition technique to provide good quality thin films with control over the chemical composition. ALD uses sequential pulses of precursors which react with the surface in a self-limiting way [12,13], giving better control of material thickness, stoichiometry and conformality than normally afforded via CVD or PVD [14,15].

Careful consideration must be exercised in precursor selection for all vapor deposition processes. Whilst they may be gas, liquid or solid at room temperature, they must be volatile and thermally stable under zationvaporization conditions. Widely used oxygen-containing precursors are H_2O, O_2 and O_3, whilst precursors for metals include metal halides, metal alkyls, metal alkoxides, etc [16]. Typical precursors for Al include aluminum chloride $[AlCl_3]$ [17], trimethylaluminum [TMA, $Al(CH_3)_3$] [18], aluminum tri-isopropoxide [AIP, $Al(O^iPr)_3$] [19] and aluminum acetylacetonate $[Al(acac)_3]$ [20]. Among the above, TMA is the most commonly used Al precursor, but it is highly reactive, moisture sensitive and pyrophoric, which makes handling inconvenient. The most common non-pyrophoric Al precursor is Al(O^{i}Pr)$_3$, but as a solid precursor there is significant possibility of condenzation during vapor transport leading to blockages and consequent reactor downtime. Therefore, in this study, the suitability of aluminum tri-sec-butoxide (ATSB, [Al(O^{s}Bu)$_3$], [Al(OCH(CH$_3$)CH$_2$CH$_3$)$_3$] was investigated as an Al precursor. At room temperature, ATSB is a non-pyrophoric liquid which contains no halogens atoms (Figure 1). ATSB has been previously reported as a metal-organic chemical vapor deposition (MOCVD) precursor [21]. However, to the best of our knowledge, it has never been studied for use in ALD before.

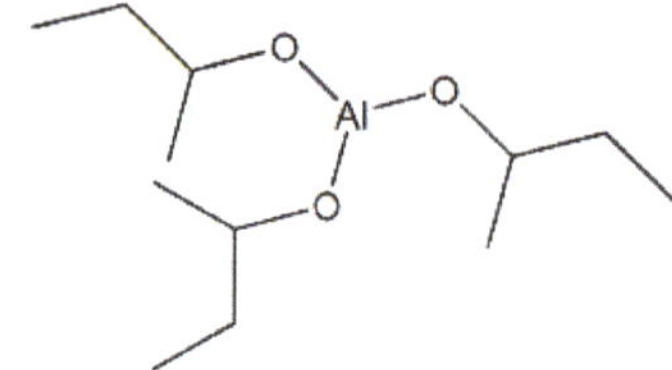

Figure 1. The structural formulas of aluminum tri-sec-butoxide.

In this research, ATSB was used in single-source CVD (also in 'pulsed CVD' mode) and ALD to deposit Al$_2$O$_3$ thin films. The deposition parameters (precursor temperature, deposition temperature, carrier gas flow rate, pulse time and purge time) were optimized on different substrates to demonstrate controllable and uniform film deposition.

2. Materials and Methods

2.1. Chemicals and Materials

Aluminum tri-sec-butoxide (ATSB, 97%) was purchased from Sigma-Aldrich (St. Louis, MI, USA). Propan-2-ol (laboratory reagent grade), methanol (HPLC grade) and ethanol absolute (HPLC grade) were received from Fisher Scientific (Loughborough, UK). Super premium glass microscope slides were purchased from VWR International, LLC (Radnor, PA, USA). Quartz slides were obtained from Wuxi Crystal and Optical Instrument Co., Ltd (Wuxi, China). P-doped silicon wafers were bought from Suzhou Crystal Silicon Electronics & Technology Co., Ltd (Suzhou, China). Deionized water was obtained from ELGA Purelab Option (ELGA LabWater). Liquid nitrogen and argon gas were provided by the BOC group (Surrey, UK).

2.2. Al$_2$O$_3$ Thin Film Deposition via CVD

Aluminum tri-sec-butoxide (maintained at 120 °C) was used as a single-source CVD precursor. ATSB vapor was carried to the reaction chamber via Ar carrier gas. To operate in CVD mode, the inlet valve and outlet valve of the ATSB bubbler were kept open for the entire deposition and no water co-reactant was used. The ATSB bubbler was heated to 115 °C, 120 °C, 125 °C, respectively. The gas flow rate was set to 150 sccm. The substrate was heated to 300 °C, 350 °C or 400 °C for film deposition. Each deposition took 24 hours. A deposition mode called 'pulsed CVD' was used to mimic the metal precursor half cycle in ALD. A cycle in pulsed CVD mode only contained one ATSB pulse (20 s or 1 min) and one argon purge (1 min). For all experiments, the pipework was heated to 170 °C to prevent precursor condenzation.

2.3. Al$_2$O$_3$ Thin Film Deposition via ALD

Al$_2$O$_3$ films were grown via ALD between 250 and 1000 deposition cycles. ATSB (maintained at 120 °C) and deionized water (maintained at 5 °C) were held in bubblers as precursors and the precursor vapor was carried to the reaction chamber via Ar carrier gas. Each cycle consisted of an ATSB pulse (2.5–20 s), an argon purge (1–3 min), then followed by a 2 s water vapor pulse and an argon purge of at least 3 min. It is worth noting that the minimum dose time in this reactor is 2 s, which inevitably requires longer purge times than normally observed in an ALD reaction. The gas flow rate through the ATSB bubbler and through the purge line was set to 120 sccm. The substrate was heated to 200 °C, 250 °C, 300 °C or 350 °C for film deposition, respectively. Glass, quartz and silicon substrates were used for the comparative study of film growth and quality. Al_2O_3 films of different thickness (20–300 cycles) were also deposited on Au NPs/WO_3 NRs system to demonstrate uniformity.

2.4. Characterization

The thickness of the films was measured using a Variable Angle Spectroscopic Ellipsometer (VASE, SEMILAB SE-2000, Semilab, Budapest, Hungary) in the wavelength range of 1.25–5 eV with variable measuring angle (56°, 57°, 58°, 59° and 60°). The optical model used for ellipsometric characterization of the ALD-like thin films was an effective medium approximation (EMA) comprising of air and the aluminum oxide (previously reported refractive index) with variable concentration [22,23]. Glass substrates were modelled using standard parameters for sodlime glass and the silicon substrates were assumed to have a 1 nm native oxide. Fittings of the ellipsometric data converged without the need for substrate-film and film-air diffusion layers. The crystalline structure of the samples was examined by X-ray diffraction (XRD, Bruker, Billerica, MA, USA, D8 Discover LynxEye) with a current of 40 mA and a voltage of 40 kV. The data was collected with a scanning rate of 0.05°/s over a 2θ range from 10° to 66° counted at 2 s per step. X-ray photoelectron spectroscopy (XPS, Thermo Fisher Scientific, Waltham, MA, USA, Neslab ThermoFlex1400) with a monochromatic Al Kα X-ray source was used to measure the elemental composition and electronic states of the elements. Tapping mode atomic force microscopy (AFM, Bruker, Billerica, MA, USA, Dimension FastScan2-SYS) using a NCLR tip was employed to investigate the surface structure of the sample. The morphologies of the samples were observed by field emission scanning electron microscope (FESEM, JEOL, Tokyo, Japan, JSM-7800F) with a current of 10 mA and an acceleration voltage of 5–15 kV after Au coating. Films deposited on Au/WO_3 films and were examined by transmission electron microscopy (TEM, JEOL, Tokyo, Japan, JEM-2100) after they were sonicated in methanol and dropped on copper grids to show the details. Photoelectrochemical (PEC) measurement was carried out using a homebuilt PEC system with a Xe lamp (75W, USHIO, California, USA) equipped with an AM 1.5 G filter (Newport, California, USA) using a Pt mesh counter electrode, a Ag/AgCl reference electrode and Na_2SO_4 electrolyte. All the photoelectrochemical tests were performed using a potentiostat (Interface 1000, Gamry, Pennsylvania, USA).

3. Results and Discussion

3.1. Deposition Parameter Optimization for CVD

Aluminum sec-butoxide ($[Al(O^sBu)_3]$) is a high viscosity liquid metal alkoxide, typically preferred for CVD reactions in comparison to solid precursors such as $Al(O^iPr)_3$. The vapor pressure of ATSB at different temperatures was calculated based on data from literature [24,25] (Figure S1). Note that the function of vapor pressure in the Antoine equation is fitted as $lg(PTorr) = 10.16 - 4177.25/T$.

According to our previous study [26], a metal precursor vapor pressure higher than 0.13 Torr would generate enough metal precursor vapor in our reactor for one pulse. Therefore, a temperature above 115 °C would provide sufficient vapor pressure.

Initial experiments focused on optimizing the precursor temperature, gas flow rate and deposition temperature for CVD deposition. After several attempts at 115 °C, 120 °C and 125 °C, the optimum ATSB bubbler temperature was found to be 120 °C. Meanwhile, increasing the precursor temperature further did not produce any significant increase in thickness at a deposition temperature of 350 °C (additional precursor not utilised in deposition). The composition of the films was analyzed by XRD, Raman spectroscopy (inVia Raman microscope, Renishaw, Gloucestershire, UK) and XPS. The crystallization temperature of γ-Al_2O_3 is at least 600 °C and 1000 °C is needed for it to transform to α-Al_2O_3 [27–29]. As the deposition temperature here was much lower than the crystallization temperature of Al_2O_3, it is consistent that neither XRD patterns nor Raman spectra showed pronounced peaks (Figures S2 and S3). For a film deposited at 400 °C for 24 h with 120 °C ATSB precursor temperature and 150 sccm gas flow, the Al 2p spectrum (Figure 2a) displayed an intense peak for the Al 2p ionization at 73.6 eV which matches with the binding energy of Al^{3+} in Al_2O_3 [30]. The Si 2p peak is of very low intensity (Figure 2b), which indicates that no Si from the glass substrate is observed, thus the Al_2O_3 film covers the entire area of the analysis spot. These findings are consistent with the films deposited via CVD using ATSB as a single-source precursor as an amorphous Al_2O_3.

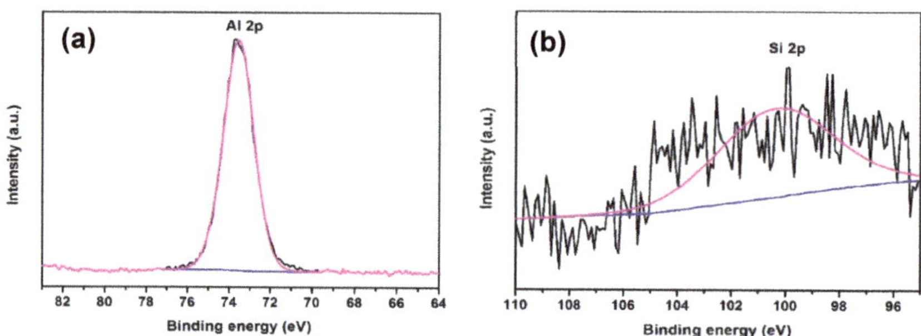

Figure 2. High resolution XPS spectra of (**a**) Al 2p and (**b**) Si 2p of Al_2O_3 film deposited on glass via CVD.

To observe the effect of substrate temperature, the CVD deposition temperatures were set to 300 °C, 350 °C and 400 °C (with a bubbler temperature of 125 °C). AFM scans (Figure 3) show an increase in the roughness of film surface as deposition temperature increases; $R_{q\,(300\,°C)}$ = 3.45 nm, $R_{q\,(350\,°C)}$ = 3.76 nm and $R_{q\,(400\,°C)}$ = 18.21 nm, which can also be seen in SEM images (Figure 3). Increasing the deposition temperature increases film roughness, which is typically unfavorable for protective coatings. Consequently, we concluded that 350 °C provided a good balance between growth rate and surface roughness.

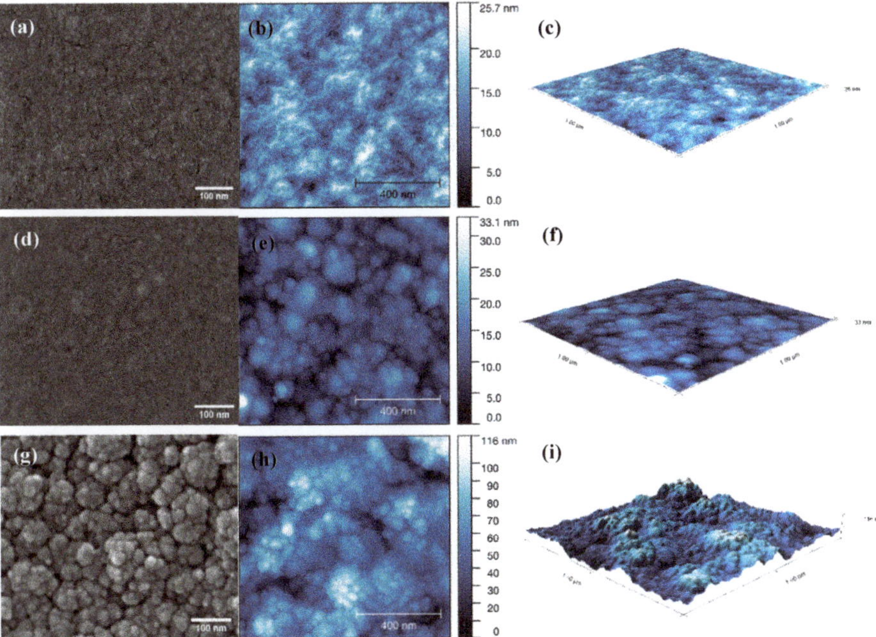

Figure 3. SEM images, AFM scans and 3D images of Al_2O_3 films deposited for 24 h with 150 sccm Ar gas flow, precursor temperature at 120 °C and deposition temperature at (**a–c**) 300 °C; (**d–f**) 350 °C; (**g–i**) 400 °C.

3.2. Pulsed CVD

After the optimization of CVD parameters, a pulsed CVD process was introduced to identify the minimum duration required for a metal precursor pulse, thereby preventing over-dosing. In order to obtain vapor transport (film growth), the pulse of ATSB was set to 20 s or 1 min, the deposition temperature was 350 °C and the precursor temperature was fixed at 120 °C with an Ar flow rate of 120 sccm. Comparing the AFM images of Al_2O_3 films deposited via continuous CVD (24 h), pulse CVD (1 min ATSB pulse, 1 min Ar purge, 1000 cycles) and pulsed CVD (20 s ATSB pulse, 1 min Ar purge, 1000 cycles). Figure 4 shows the surface roughness decrease from $R_{q\,(continuous\;deposition)} = 10.19$ nm to $R_{q\,(1min\;pulse)} = 2.51$ nm and $R_{q\,(20\;s\;pulse)} = 2.27$ nm. We concluded from these data that a 20 s pulse produced complete coverage and was therefore a suitable starting point for ATSB precursor dosing in an ALD process.

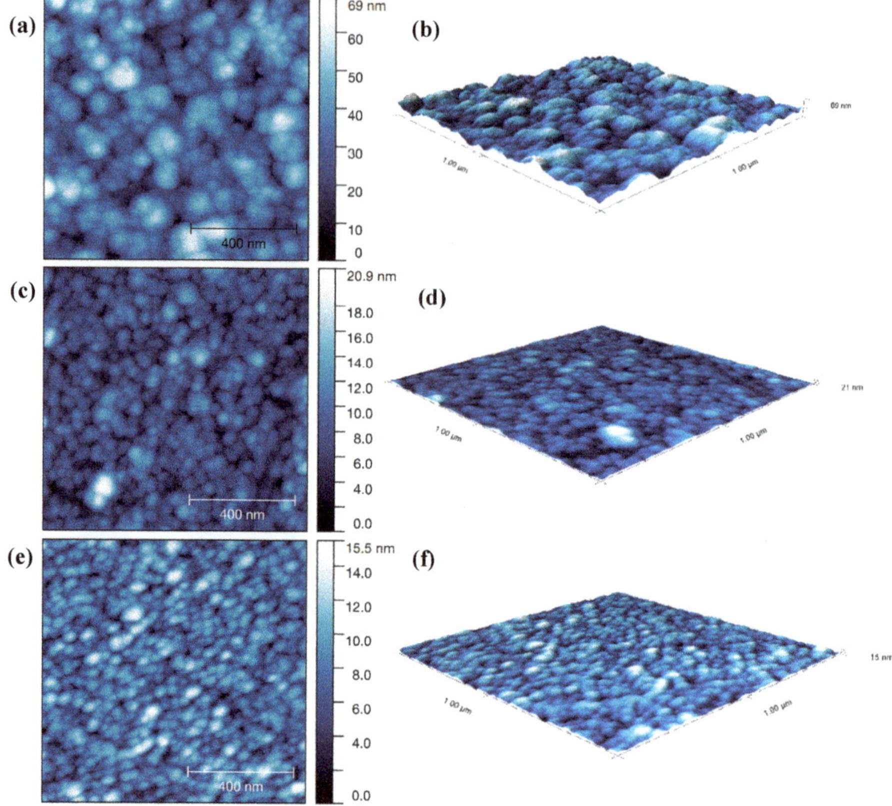

Figure 4. AFM scans and 3D images of Al$_2$O$_3$ films deposited at 350°C via (**a,b**) continuous CVD; (**c,d**) pulsed CVD (1 min ATSB pulse) and (**e,f**) pulsed CVD (20 s ATSB pulse).

3.3. Deposition Parameter Optimization for ALD

The reaction mechanism when using ATSB and water as ALD precursors may be expected as [31,32]:

$$x \text{ -OH (s)} + \text{Al(O}^s\text{Bu)}_3 \text{ (g)} \rightarrow \text{(-O-)}_x\text{Al(O}^s\text{Bu)}_{3-x} \text{ (s)} + x \, ^s\text{BuOH (g)} \tag{1}$$

$$\text{(-O-)}_x\text{Al(O}^s\text{Bu)}_{3-x} \text{ (s)} + (3-x) \, \text{H}_2\text{O (g)} \rightarrow \text{(-O-)}_x\text{Al(OH)}_{3-x} \text{ (s)} + (3-x) \, ^s\text{BuOH (g)} \tag{2}$$

with $x = 1$ being the monodentate, $x = 2$, bidentate and $x = 3$, tridentate, respectively, depending on the number of alkoxide ligands.

A full ALD reaction contains two half reactions. In the first half reaction, ATSB reacts with OH groups on the substrate surface to produce $(–\text{O}–)_x\text{Al(O}^s\text{Bu)}_{3-x}$, where x is the number of OH surface sites react with ATSB which also determines the geometrical configuration on the surface. The $(–\text{O}–)_x\text{Al(O}^s\text{Bu)}_{3-x}$ group can react with H$_2$O to form $(–\text{O}–)_x\text{Al(OH)}_{3-x}$ in the second half of the reaction, therefore, the surface layer is covered with OH groups again, allowing the first half reaction to take place again [33].

The deposition temperature used in ALD should be lower than that used in CVD to avoid uncontrolled CVD-like film growth, therefore the ALD deposition temperature was decreased in comparison to CVD in increments from 350 °C to 300 °C, 250 °C and 200 °C whilst the other parameters were kept unchanged (20 s ATSB pulse, 1 min Ar purge, 2 s H$_2$O pulse and 3 min Ar purge with 120 sccm Ar gas flow deposited for 500 cycles). XPS spectra of this series of samples indicate that films deposited at 200 °C, 250 °C and 300 °C show similar binding energy intensity at approximately 74 eV (Figure 5). The growth rate of Al_2O_3 was between 1.9–2.3 nm/cycle in the deposition temperature range of 200 °C–300 °C with a 20 s pulse time, thus 200 °C was selected in further experiments (Figure S4). The thickness and growth rate were found to be much higher than previously reported values (0.1 nm/cycle) [34–39] indicating a combination of ALD and CVD-like growth. One likely explanation is that the dose time of ATSB was too long for the purge time used, therefore the ATSB dose time was reduced from 20 s to 2.5 s with the aim to avoid CVD-like growth. Changing the ATSB purge time from 1 min to 3 min reduced the deposition rate, while more than 3 min for a 2.5 s dose time resulted in no change in the growth rate, therefore 3 min purge was considered optimum for a 2.5 s Similarly, for a 2 s H_2O dose, changing the purge time from 3 min to 5 min had no effect, indicating that a 3 min purge was sufficient to ensure all water was evacuated between cycles.

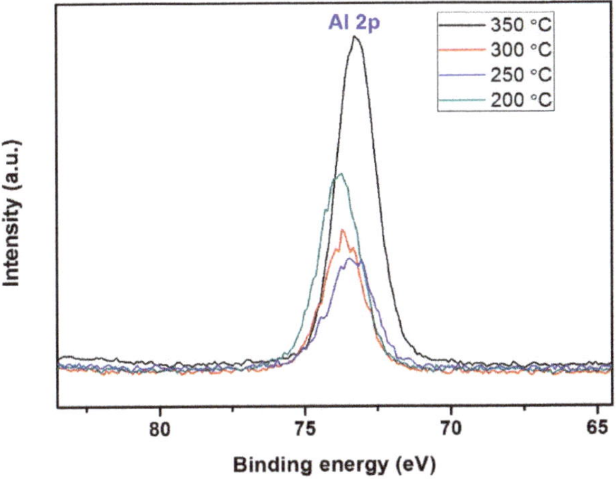

Figure 5. High resolution XPS spectra of Al elements under different deposition temperature, 200 °C, 250 °C, 300 °C and 350 °C.

With a 2.5 s dose time, the thickness reduced to approx. 50 nm after 500 cycles, which gave growth rates in the range of 0.12 to 0.15 nm/cycle with successful deposition on silicon, quartz and glass substrates (Table S1). The growth rate using ATSB as an ALD precursor is similar to that using TMA or $AlCl_3$ (0.08–0.2 nm/cycle) [38,40–42]. Thus, the ALD-like parameters were fixed at an ATSB bubbler temperature of 120 °C, deposition temperature of 200 °C and gas flow rate of 120 sccm (a full cycle includes 2.5 s ATSB pulse, 3 min Ar purge, 2 s H_2O pulse and 3 min Ar purge).

The density of the Al_2O_3 component in the ellipsometric EMA model varied between 0.720 and 0.757 (Table S1 and Figure S5). The films deposited with short 2.5 s ATSB pulse times showed lower growth rates (0.12–0.15 nm/cycle) and higher densities than those grown with longer 20 s pulses. The root mean square roughness of Al_2O_3 film on three substrates were $R_{q(silicon)}$ = 1.48 nm, $R_{q(quartz)}$ = 1.16 nm and $R_{q(glass)}$ = 0.96 nm, suggesting an applicability of Al_2O_3 on a variety of substrates (Figure 6).

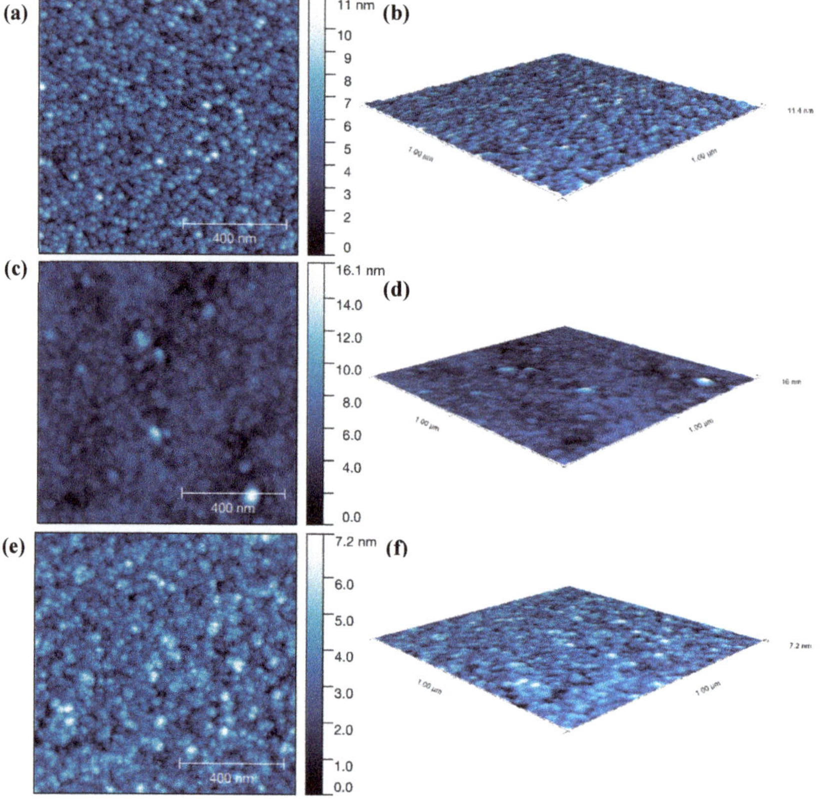

Figure 6. AFM scans and 3D images of Al_2O_3 films after 250 ALD cycles, on (**a,b**) silicon; (**c,d**) quartz; (**e,f**) glass. A cycle includes 2.5 s ATSB pulse, 3 min Ar purge, 2 s H_2O pulse and 3 min Ar purge, deposition temperature 200 °C, gas flow rate 120 sccm.

3.4. ALD Al_2O_3 Films on Complex Structured Au/WO$_3$

Our aim is to use this ALD Al_2O_3 film as a passivation layer for photoelectrode. In former studies of ALD thin films that include Al_2O_3 and other oxides [43–45], it has been mentioned that either an ultrathin film of several nanometers or a thin film with a thickness above 20 nm showed different degrees of current density enhancement during a photoelectrochemical test.

Al_2O_3 films were deposited via the ALD-like process on a complex nanostructure consisting of Au nanoparticles on needle-like WO_3 nanostructures (Au/WO$_3$, Figures 7 and 8). Au/WO$_3$ is an electrode prepared in house on an FTO glass for water splitting. However, it is not stable in any kind of electrolyte. Therefore, Al_2O_3 films were deposited with the aim to protect the electrode without severely affecting the photocurrent required for water splitting. The SEM image of the Au/WO$_3$ shows a plane of nanoneedles which has a high surface area and high porosity. Due to the complex structure, the thickness of the ALD layer was difficult to measure using an optical method, therefore film thickness during deposition was recorded using a 'witness' planar substrate inserted in the reactor at the same time with the sample. Growth rate as a function of number of cycles was plotted (Figure 8a). The growth rate of Al_2O_3 on both the witness plate and the nanostructures was between 0.1 nm/cycle and 0.13 nm/cycle (shown in Figure S6). For a given number of cycles, similar thickness were achieved on both surfaces. There is a strong correlation between the film thickness and the number of ALD cycles (Table 1), e.g., 50 cycles and 300 cycles gave 4 nm and 40 nm respectively on both the planar substrate (measured using ellipsometry) and the complex nanomaterial architecture (measured by TEM, Figure 8c,d).

Figure 7. SEM image of Au/WO$_3$ film deposited via CVD (top view).

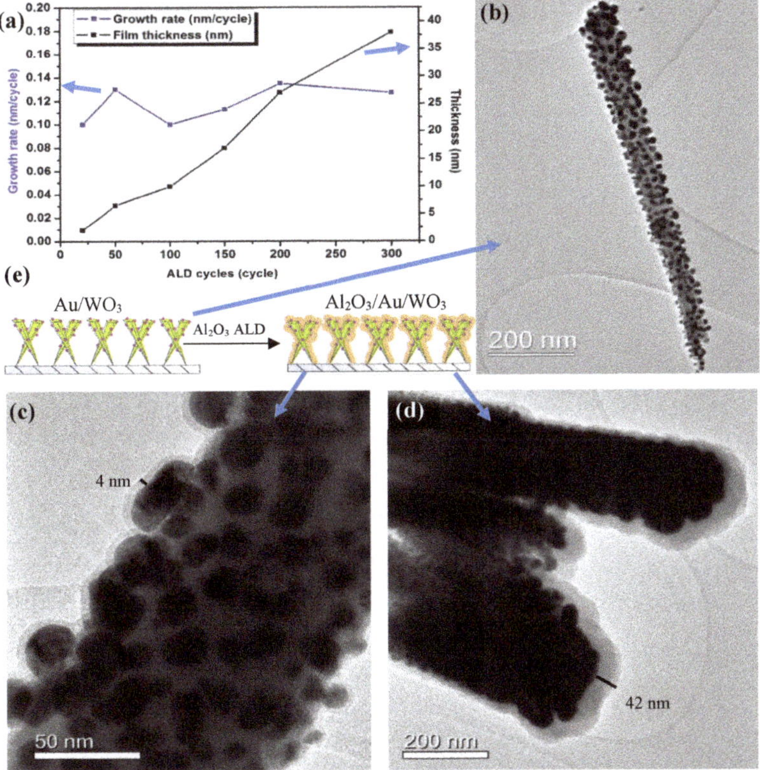

Figure 8. (**a**) Al$_2$O$_3$ film growth rate (blue) and film thickness (black) as a function of reaction cycles, (**b**) Au/WO$_3$ structure, (**c**) Au/WO$_3$ with 50 cycles ALD Al$_2$O$_3$ growth on top, (**d**) Au/WO$_3$ with 300 cycles ALD Al$_2$O$_3$ growth on top, (**e**) Schematic procedure of ALD Al$_2$O$_3$ growth.

Table 1. Summary of the ALD Al_2O_3 thin film deposition parameters and the film thickness on Au/WO_3, FTO and silicon.

Substrate	Deposition Recipe				Cycle (Number)	Substrate Temperature (°C)	Flow Rate (sccm)	Thickness (nm)
	ATSB	Purge	H_2O	Purge				
Silicon, FTO and Au/WO_3	2.5 s	3 min	2 s	3 min	20	200	120	2
					50			7
					100			10
					150			15
					200			20
					300			40

The XPS spectrum of Al 2p shows that the intensity of Al 2p peak increases with the increase of the number of cycles (Figure 9a). In contrast, the intensity of Au 4f peak decreases with the increase of the number of cycles, with very little Au 4f signal intensity remaining after 50 cycles (~6 nm Al_2O_3) and no Au signal detected after 150 cycles (~18 nm) (Figure 9b). Our interest in the conformal Al_2O_3 is as a barrier/corrosion resistant layer for electrochemistry/photoelectrochemistry. To evaluate the quality of the coating, the photoelectrochemical performance was tested (Figure 10). The results demonstrate that increasing the Al_2O_3 film thickness beyond 50 deposition cycles (~6 nm) lead to a significant decrease in photocurrent density, as expected for high dieletric material. Consequently, both the XPS and PEC data indicate that even for a 50 cycles coating, the Al_2O_3 film was conformal and without pinholes even on these high aspect ratio, porous nanostructures.

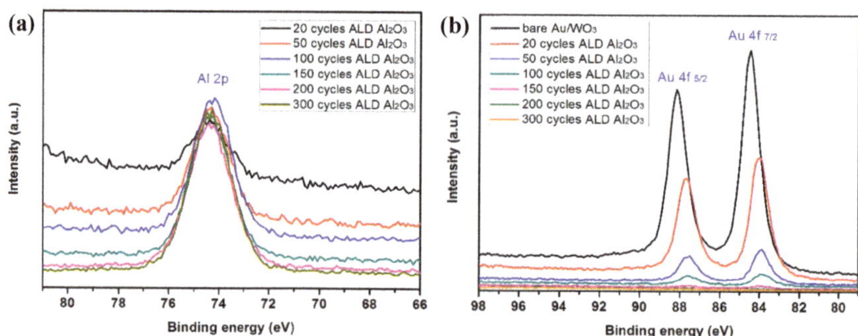

Figure 9. XPS spectra of (**a**) Al 2p and (**b**) Au 4f in $Al_2O_3/Au/WO_3$ nanostructures with various cycles of ALD Al_2O_3 thin film.

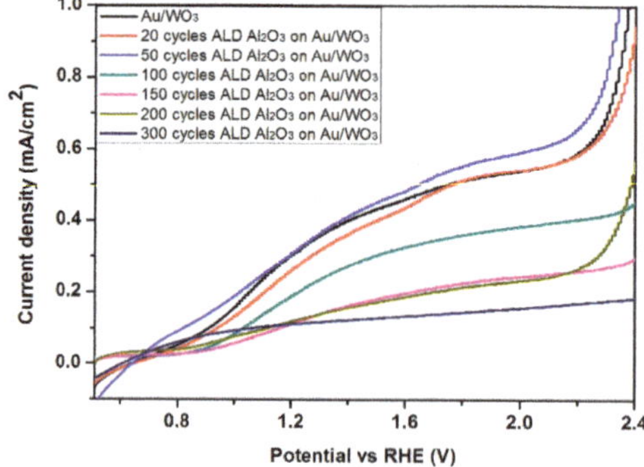

Figure 10. J-V curves of $Al_2O_3/Au/WO_3$ during photoelectrochemical measurement.

4. Conclusions

In the study, the parameters of Al_2O_3 thin films atomic layer deposition using aluminum tri-sec-butoxide as a new Al precursor and water for ALD were investigated. The CVD, pulsed CVD and ALD process was optimized, with the best deposition conditions for ALD as follows: ATSB bubbler temperature 120 °C; deposition temperature 200 °C; gas flow rate 120 sccm; ATSB pulse time 2.5 s; Ar purge time 3 min; H_2O pulse time 2 s; Ar purge time 3 min. A stable ALD process with a growth rate of 0.12–0.15 nm/cycle was observed by measuring the Al_2O_3 film thickness on different substrates using ellipsometry. The composition and morphology of the as-synthesized films were analyzed and followed by the comparison of film deposition under different conditions. SEM, AFM and XPS data indicated that the obtained films were dense and continuous with a low concentration of impurities. TEM images of Al_2O_3 thin films prepared from ATSB and water on complex nanostructures show uniformity, conformality and good control of thickness, strongly suggesting the potential of using this new ALD precursor for Al_2O_3 thin films. These Al_2O_3 films are likely to be used as a protection layer on the Au/WO_3 electrode for photoelectrochemical water splitting.

Supplementary Materials: The following are available online at http://www.mdpi.com/1996-1944/12/9/1429/s1, Figure S1: The relationship between ATSB vapor pressure and ATSB precursor temperature. Figure S2: XRD pattern of Al_2O_3 film deposited via CVD. This example was made under condition: precursor temperature 120 °C; deposition temperature 350 °C; gas flow rate 150 sccm; deposition time 24 h. Figure S3: Raman spectra of Al_2O_3 film deposited via CVD. This example was made under condition: precursor temperature 120 °C; deposition temperature 350 °C; gas flow rate 150 sccm; deposition time 24 h. Figure S4: Growth rates of the Al_2O_3 films as a function of deposition temperature from 200 °C to 300 °C established via ellipsometry. The deposition conditions were 20 s ATSB pulse, 1 min Ar purge, 2 s H_2O pulse and 3 min Ar pulse for 500 cycles. Figure S5: Refractive index of (a) Al_2O_3 [23], and (b) Al_2O_3/air (0.757/0.243) Effective Medium Approximation used for ellipsometric modelling of ALD film on glass, Figure 6e. Figure S6: Film thicknesses established via Ellipsometry for ALD growth on Silicon samples (R^2 fit quality for each ellipsometric fitting labelled). EMA concentration for all samples held at 74.5%. ATSB pulse duration 2.5 s. Table S1: Table of structural parameters from for ellipsometric model fitting between (1.25–5 eV) for samples produced via ALD (250 cycles: ATSB pulse indicated, 3 min purge, 2 s H_2O pulse, 3 min purge).

Author Contributions: Conceptualization, X.X. and C.B.; Data curation, X.X.; Formal analysis, X.X., A.T.; Funding acquisition, C.B.; Methodology, X.X. and S.G.; Project administration, C.B. and S.G.; Resources, X.X., A.T., S.G. and C.B.; Software, X.X., A.T. and Y.Z.; Supervision, C.B. and S.G.; Visualization, X.X. and Y.Z.; Writing —original draft, X.X.; Writing—review & editing, X.X., S.G., A.T., Y.Z. and C.B.

Funding: This research received no external funding.

Acknowledgments: The authors gratefully acknowledge the support by Department of Chemistry, University College London and China Scholarship Council. The authors would also like to acknowledge support from Rachel Wilson, Yaomin Li, Steven Firth, Robert Palgrave and Steve Hudziak.

Conflicts of Interest: The authors declare no conflict of interest.

References

1. Bordihn, S.; Engelhart, P.; Mertens, V.; Kesser, G. High surface passivation quality and thermal stability of ALD Al_2O_3 on wet chemical grown ultra-thin SiO_2 on silicon. *Energy Procedia* **2011**, *8*, 654–659. [CrossRef]
2. Dingemans, G.; Seguin, R.; Engelhart, P.; Van De Sanden, M.C.M.; Kessels, W.M.M. Silicon surface passivation by ultrathin Al_2O_3 films synthesized by thermal and plasma atomic layer deposition. *Phys. Status Solidi (RRL)* **2010**, *4*, 10–12. [CrossRef]
3. Tang, S.H.; Kuo, C.I.; Trinh, H.D.; Hudait, M.; Chang, E.Y.; Hsu, C.Y.; Su, Y.H.; Luo, G.L.; Nguyen, H.Q. C–V characteristics of epitaxial germanium metal-oxide-semiconductor capacitor on GaAs substrate with ALD Al_2O_3 dielectric. *Microelectron. Eng.* **2012**, *97*, 16–19. [CrossRef]
4. Ritala, M.; Leskela, M.; Dekker, J.-P.; Mutsaers, C.; Soininen, P.J.; Skarp, J. Perfectly conformal TiN and Al_2O_3 films deposited by atomic layer deposition. *Chem. Vap. Depos.* **1999**, *5*, 7–9. [CrossRef]
5. Schmidt, J.; Merkle, A.; Brendel, R.; Hoex, B.; Van De Sanden, M.C.M.; Kessels, W.M.M. Surface passivation of high-efficiency silicon solar cells by atomic-layer-deposited Al_2O_3. *Prog. Photovoltaics Res. Appl.* **2008**, *16*, 461–466. [CrossRef]

6. Siol, S.; Hellmann, J.C.; Tilley, S.D.; Graetzel, M.; Morasch, J.; Deuermeier, J.; Jaegermann, W.; Klein, A. Band alignment engineering at Cu_2O/ZnO heterointerfaces. *ACS Appl. Mater. Interfaces* **2016**, *8*, 21824–21831. [CrossRef] [PubMed]

7. García-valenzuela, J.A.; Rivera, R.; Morales-vilches, A.B.; Gerling, L.G.; Caballero, A.; Asensi, J.M.; Voz, C.; Bertomeu, J.; Andreu, J. Main properties of Al_2O_3 thin films deposited by magnetron sputtering of an Al_2O_3 ceramic target at different radio-frequency power and argon pressure and their passivation effect on p-type c-Si wafers. *Thin Solid Films* **2016**, *619*, 288–296. [CrossRef]

8. Hoffman, D.; Leibowitz, D. Al_2O_3 films prepared by electron-beam evaporation of hot-pressed Al_2O_3 in oxygen ambient. *J. Vac. Sci. Technol.* **1971**, *8*, 107–111. [CrossRef]

9. Landälv, L. Thin film and plasma characterization of PVD oxides. Ph.D. Thesis, Linköping University, Linköping, Sweden, 2017.

10. Mcclure, C.D.; Oldham, C.J.; Parsons, G.N. Surface & coatings technology effect of Al_2O_3 ALD coating and vapor infusion on the bulk mechanical response of elastic and viscoelastic polymers. *Surf. Coat. Technol.* **2015**, *261*, 411–417. [CrossRef]

11. Yang, P.; Yao, M.; Xiao, R.; Chen, J.; Yao, X. Preparation of Al_2O_3 film by sol-gel method on thermally evaporated Al film. *Vacuum* **2014**, *107*, 112–115. [CrossRef]

12. George, S.M. Atomic layer deposition: an overview. *Chem. Rev.* **2009**, *110*, 111–131. [CrossRef] [PubMed]

13. Puurunen, R.L. A short history of atomic layer deposition: Tuomo Suntola's atomic layer epitaxy. *Chem. Vapor Depos.* **2014**, *20*, 332–344. [CrossRef]

14. Jones, A.C.; Hitchman, M.L. *Chemical Vapour Deposition: Precursors, Processes and Applications*; Royal Society of Chemistry: Cambridge, UK, 2009.

15. Johnson, R.W.; Hultqvist, A.; Bent, S.F. A brief review of atomic layer deposition: from fundamentals to applications. *Mater. Today* **2014**, *17*, 236–246. [CrossRef]

16. Multone, X.; Hoffmann, P. *High Vacuum Chemical Vapor Deposition (HV-CVD) of Alumina Thin Films*; École polytechnique fédérale de Lausanne: Lausanne, Switzerland, 2009.

17. Muhsin, A.E. Chemical vapor deposition of aluminum oxide (Al_2O_3) and beta iron disilicide (β-$FeSi_2$) thin films. Ph.D. Thesis, University of Duisburg-Essen, Duisburg, Germany, 2007.

18. Liu, X.; Chan, S.H.; Wu, F.; Li, Y.; Keller, S.; Speck, J.S.; Mishra, U.K. Metalorganic chemical vapor deposition of Al_2O_3 using trimethylaluminum and O_2 precursors: growth mechanism and crystallinity. *J. Cryst. Growth.* **2014**, *408*, 78–84. [CrossRef]

19. Blittersdorf, B.S.; Bahlawane, N.; Kohse-höinghaus, K.; Atakan, B.; Müller, J. CVD of Al_2O_3 Thin films using aluminum tri-isopropoxide. *Chem. Vap. Depos.* **2003**, *4*, 194–198. [CrossRef]

20. Ponja, S.D.; Parkin, I.P.; Carmalt, C.J. Synthesis and material characterization of amorphous and crystalline (α-)Al_2O_3 via aerosol assisted chemical vapour deposition. *RSC Adv.* **2016**, *6*, 102956–102960. [CrossRef]

21. Haanappel, V.A.; Van Corbach, H.D.; Fransen, T.; Gellings, P.J. The pyrolytic decomposition of aluminium-tri-sec-butoxide during chemical vapour deposition of thin alumina films. *Thermochim. Acta.* **1994**, *240*, 67–77. [CrossRef]

22. Tompkins, H.G.; Irene, E.A. (Eds.) *Handbook of Ellipsometry*; Irene Publisher Springer: New York, NY, USA, 2010.

23. Palik, E.D. *Handbook of Optical Constants of Solids*; Academic Press: Palm Bay, FL, USA, 1997.

24. Aluminium tri-sec-butoxide(2269-22-9) MSDS Melting Point Boiling Point Density Storage Transport. Available online: http://www.chemicalbook.com/ProductMSDSDetailCB5397907_EN.htm (accessed on 13 January 2018).

25. Aluminium tri-sec-butoxide SAFETY DATA SHEET. Available online: https://www.fishersci.com/shop/msdsproxy?productName=AC191215000&productDescription=ALUMINUM+TRI-SEC-BUTOXID+500GR&catNo=AC191215000&vendorId=VN00032119&storeId=10 (accessed on 13 January 2018).

26. Wilson, R.; Simion, C.; Blackman, C.; Carmalt, C.; Stanoiu, A.; Di Maggio, F.; Covington, J. The effect of film thickness on the gas sensing properties of ultra-thin TiO_2 films deposited by atomic layer deposition. *Sensors* **2018**, *18*, 735. [CrossRef]

27. Sathyaseelan, B.; Baskaran, I.; Sivakumar, K. Phase transition behavior of nanocrystalline Al_2O_3 powders. *Soft Nanosci. Let.* **2013**, *3*, 69. [CrossRef]

28. Prokes, S.M.; Katz, M.B.; Twigg, M.E. Growth of crystalline Al_2O_3 via thermal atomic layer deposition: nanomaterial phase stabilization. *APL Mater.* **2014**, *2*, 032105. [CrossRef]

29. Cava, S.; Tebcherani, S.M.; Souza, I.A.; Pianaro, S.A.; Paskocimas, C.A.; Longo, E.; Varela, J.A. Structural characterization of phase transition of Al_2O_3 nanopowders obtained by polymeric precursor method. *Mater. Chem. Phys.* **2007**, *103*, 394–399. [CrossRef]

30. NIST XPS Database Detail Page. Available online: https://srdata.nist.gov/xps/XPSDetailPage.aspx?AllDataNo=67580 (accessed on 15 January 2018).

31. Guerra-Nuñez, C.; Döbeli, M.; Michler, J.; Utke, I. Reaction and growth mechanisms in Al_2O_3 deposited via atomic layer deposition: elucidating the hydrogen source. *Chem. Mater.* **2017**, *29*, 8690–8703. [CrossRef]

32. Juppo, M.; Rahtu, A.; Ritala, M.; Leskelä, M. In situ mass spectrometry study on surface reactions in atomic layer deposition of Al_2O_3 thin films from trimethylaluminum and water. *Langmuir* **2000**, *16*, 4034–4039. [CrossRef]

33. Matero, R.; Rahtu, A.; Ritala, M.; Leskelä, M.; Sajavaara, T. Effect of water dose on the atomic layer deposition rate of oxide thin films. *Thin Solid Films* **2000**, *368*, 1–7. [CrossRef]

34. Barbos, C.; Blanc-pelissier, D.; Fave, A.; Balnquet, E.; Crisci, A.; Fourmond, E.; Albertini, D.; Sabac, A.; Ayadi, K.; Girard, P.; Lemiti, M. Characterization of Al_2O_3 thin films prepared by thermal ALD. *Energy Procedia* **2015**, *77*, 558–564. [CrossRef]

35. Barbos, C.; Blanc-pelissier, D.; Fave, A.; Botella, C.; Regreny, P.; Grenet, G.; Blanquet, E.; Crisci, A.; Lemiti, M. Al_2O_3 thin films deposited by thermal atomic layer deposition: characterization for photovoltaic applications. *Thin Solid Films* **2016**, *617*, 108–113. [CrossRef]

36. Batra, N.; Gope, J.; Panigrahi, J.; Singh, R.; Singh, P.K. Influence of deposition temperature of thermal ALD deposited Al_2O_3 films on silicon surface passivation. *Aip Adv.* **2015**, *5*, 067113. [CrossRef]

37. Campabadal, F.; Beldarrain, O.; Zabala, M.; Acero, M.C.; Rafí, J.M. Comparison between Al_2O_3 thin films grown by ALD using H_2O or O_3 as oxidant source. In Proceedings of the 8th Spanish Conference on Electron Devices, Palma de Mallorca, Spain, 8–11 February 2011.

38. Groner, M.D.; Fabreguette, F.H.; Elam, J.W.; George, S.M. Low-temperature Al_2O_3 atomic layer deposition. *Chem. Mater.* **2004**, *16*, 639–645. [CrossRef]

39. Vähä-nissi, M.; Sievänen, J.; Salo, E.; Heikkilä, P.; Kenttä, E.; Johansson, L.-S.; Koskinen, J.T.; Harlin, A. Atomic and molecular layer deposition for surface modification. *J. Solid State Chem.* **2014**, *214*, 7–11. [CrossRef]

40. Räisänen, P.I.; Ritala, M.; Leskelä, M. Atomic layer deposition of Al_2O_3 films using $AlCl_3$ and $Al(O^iPr)_3$ as precursors. *J. Mater. Chem.* **2002**, *12*, 1415–1418. [CrossRef]

41. Kim, J.W.; Kim, D.H.; Oh, D.Y.; Lee, H.; Kim, J.H.; Lee, J.H.; Jung, Y.S. Surface chemistry of $LiN_{i0.5}Mn_{1.5}O_4$ particles coated by Al_2O_3 using atomic layer deposition for lithium-ion batteries. *J. Power Sources* **2015**, *274*, 1254–1262. [CrossRef]

42. Iatsunskyi, I.; Kempiński, M.; Jancelewicz, M.; Załęski, K.; Jurga, S.; Smyntyna, V. Structural and XPS characterization of ALD Al_2O_3 coated porous silicon. *Vacuum* **2015**, *113*, 52–58. [CrossRef]

43. Paracchino, A.; Mathews, N.; Hisatomi, T.; Stefik, M.; Tilley, S.D.; Grätzel, M. Ultrathin films on copper (I) oxide water splitting photocathodes: a study on performance and stability. *Energy Environ. Sci.* **2012**, *5*, 8673–8681. [CrossRef]

44. Abdulagatov, A.I.; Yan, Y.; Cooper, J.R.; Zhang, Y.; Gibbs, Z.M.; Cavanagh, A.S.; George, S.M. Al_2O_3 and TiO_2 atomic layer deposition on copper for water corrosion resistance. *ACS Appl. Mater. Interfaces* **2011**, *3*, 4593–4601. [CrossRef]

45. Le Formal, F.; Tetreault, N.; Cornuz, M.; Moehl, T.; Grätzel, M.; Sivula, K. Passivating surface states on water splitting hematite photoanodes with alumina overlayers. *Chem. Sci.* **2011**, *2*, 737–743. [CrossRef]

Review

A Review of Perovskite Photovoltaic Materials' Synthesis and Applications via Chemical Vapor Deposition Method

Xia Liu [1,2,3], Lianzhen Cao [1,2,3,*], Zhen Guo [2,4,5,*], Yingde Li [1], Weibo Gao [3,*] and Lianqun Zhou [2,*]

[1] Department of Physics and Optoelectronic Engineering, Weifang University, Weifang 261061, Shandong, China; liuxia@wfu.edu.cn (X.L.); wfulyd@163.com (Y.L.)

[2] Chinese Academy of Sciences Key Lab of Bio-Medical Diagnostics, Suzhou Institute of Biomedical Engineering and Technology, Chinese Academy of Sciences, Suzhou 215163, Jiangsu, China

[3] Division of Physics and Applied Physics School of Physical and Mathematical Sciences Nanyang Technological University, Singapore 637371, Singapore

[4] Shandong Guo Ke Medical Technology Development Co., Ltd., Jinan 250001, Shandong, China

[5] Zhongke Mass Spectrometry (Tianjin) Medical Technology Co., Ltd., Tianjin 300399, China

* Correspondence: lianzhencao@wfu.edu.cn (L.C.); guozhen@sibet.ac.cn (Z.G.); wbgao@ntu.edu.sg (W.G.); zhoulq@sibet.ac.cn (L.Z.)

Received: 31 August 2019; Accepted: 8 October 2019; Published: 11 October 2019

Abstract: Perovskite photovoltaic materials (PPMs) have emerged as one of superstar object for applications in photovoltaics due to their excellent properties—such as band-gap tunability, high carrier mobility, high optical gain, astrong nonlinear response—as well as simplicity of their integration with other types of optical and electronic structures. Meanwhile, PPMS and their constructed devices still present many challenges, such as stability, repeatability, and large area fabrication methods and so on. The key issue is: how can PPMs be prepared using an effective way which most of the readers care about. Chemical vapor deposition (CVD) technology with high efficiency, controllability, and repeatability has been regarded as a cost-effective road for fabricating high quality perovskites. This paper provides an overview of the recent progress in the synthesis and application of various PPMs via the CVD method. We mainly summarize the influence of different CVD technologies and important experimental parameters (temperature, pressure, growth environment, etc.) on the stabilization, structural design, and performance optimization of PPMS and devices. Furthermore, current challenges in the synthesis and application of PPMS using the CVD method are highlighted with suggested areas for future research.

Keywords: atmospheric pressure CVD; low pressure CVD; hybrid CVD; aerosol assisted CVD; pulsed CVD; perovskite photovoltaic nanomaterials; stabilization; structural design; performance optimization; solar cells

---

1. Introduction

Perovskite is a kind of material with the same crystal structure as calcium titanate ($CaTiO_3$), which was discovered by Gustav Rose in 1839 [1]. The perovskite material (PM) structure formula is generally ABX_3, where A and B are two cations and X is anion, as shown in Figure 1 [2]. This unique crystal structure gives it many unique physical and chemical properties, such as broadband bandgap tunability, high carrier mobility, high optical gain, strong nonlinear response, and simplicity of their integration with other types of optical and electronic structures [3–8]. Due to its excellent physical and chemical properties, especially optical properties, perovskite materials are now widely applied for constructing photovoltaic solar cells (PSCs) [2,9–11].

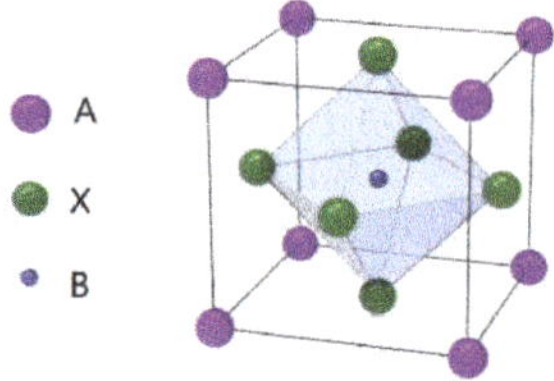

Figure 1. Crystal structure of the perovskite ABX$_3$ form.

Up to now, great progress has been made in the preparation and application of bulk PPMs. The perovskite family now includes hundreds of substances, ranging from organic materials, inorganic materials to organic–inorganic materials, from polycrystalline film to bulk single crystal [12]. However, introduction of many defects and grain boundaries in three-dimensional (3D) bulk PPMS is unavoidable [13], which inevitably degrade its optoelectronic properties. Recently, low-dimensional PPMS with high quality have begun to receive increasing attention, primarily for their tunable optical and electronic properties due to quantum-size effects as well as their enhanced photoelectric performance compared with bulk materials. Moreover, perovskites are considered to be a class of important low-dimensional layered materials, whose optical properties can be controlled by varying the number of layers, particularly at thicknesses lower than their exciton Bohr radius. Therefore, the shape-, size-, and thickness-controlled synthesis of low-dimensional PPMs has recently been a hot topic of research. Various low-dimensional morphologies, including zero-dimensional (0D) morphologies, such as quantum dots and nanoparticles; one-dimensional (1D) morphologies, such as nanowires and nanorods; and two-dimensional (2D) morphologies, such as nanoplates and nanosheets, have been prepared and applied [14–21].

Due to potential applications of perovskite materials (PMs) in photovoltaics wide spread interest has focused on exploring various synthesis methods to obtain high quality perovskite materials for constructing designed devices with improved performance. Up to now, many synthesis methods—such as solution process [4,22–26], thermal evaporation process [27], flash evaporation [28], doctor-blade coating method [29], slot-die coating method [30], spray deposition [31], ink-jet printing method [32], vapor-assisted solution process (VASP) [33], and so on—have been developed for preparing PPMs with well-designed dimensions for exploring factors which plays key role in dominating their properties. Among various approaches, the most studied solution method not only achieves highly efficient perovskite solar cells (PSCs) but also has a cost advantage. However, the uncontrolled rapid liquid reaction often generates rough, porous, and less-stable perovskite films with incomplete conversions, resulting in large variations on film morphology and Photovoltaic (PV) response [24–26]. Inversely, thermal evaporation can generate high-quality perovskite films with smooth and pinhole-free morphologies. However, there are a few of disadvantages in thermal evaporation process, such as low material utilization, and high equipment investment and energy consumption, hampering its further application [27,28]. The rest technologies often produce poor perovskite film quality, and are also difficult to scale up [29–32]. VASP methods require manipulations protecting atmosphere, and show poor controllability and versatility, unsuitable for their mass production [33]. Therefore, development of a new class of advanced fabrication technology shall be critical for the future commercialization of PPMs.

Soon after, as an evolution of VASP, an array of CVD techniques [2,34–39]—such as one-step and two-step tubular CVD, in situ tubular CVD (ITCVD), aerosol-assisted CVD (AACVD), hybrid physical-chemical vapor deposition (HPCVD), plasma-enhanced CVD (PECVD), and so on—were developed to grow high-quality PPMs. At present, CVD of perovskites is a promising alternative to solution based methods of fabrication due to the relative ease of patterning, the ability to batch process, the wide range of material compatibility, and the potential for uniform large-area deposition.

Here, in this review paper, CVD method—as one of the key technologies for preparing many kinds of advanced semiconductor materials with excellent properties—is mainly mentioned for preparing PPMs to construct optimal devices. For details, in this topical review, we present the recent developments of synthesis and application of various PPMs via the CVD method. We begin with the characteristics, classification, reaction process, and application of CVD method. In the following section, we mainly summarize the influence of different CVD technologies and important experimental parameters (temperature, pressure, growth environment, etc.) on the PPMs and devices. In particular, we aim to show why CVD technology can improve the stability of materials and devices, and can help material and device design, thus improving material and device performance. Finally, current challenges in the synthesis and application of PPMs using the CVD method are highlighted with suggested areas for future research.

2. CVD Method

CVD is a process for producing solid products from gases, and is a mature, lost cost, high efficiently technology for fabricating various kinds of semiconductor materials. CVD equipment mainly includes gas source control unit, deposition reaction chamber, deposition temperature control unit and vacuum exhaust and pressure control unit, and some experimental devices also have enhanced excitation energy control components. Therefore, according to the heating mode classification, it can be divided into thermal activation (resistance, high-frequency induction or infrared radiation heating, etc.), plasma enhancement, laser enhancement, microwave plasma enhancement, and other deposition modes, as shown in Table 1. In accordance with the reaction pressure classification, CVD can be divided into atmospheric pressure CVD (APCVD), low-pressure CVD (LPCVD) and ultrahigh vacuum CVD (UHVCVD, $<10^{-6}$ Pa). According to gas phase classification, CVD can be further divided into Metal organic chemical vapor deposition (MOCVD), AACVD, direct liquid injection CVD (DLICVD), and HPCVD [40–45].

Table 1. Different CVD devices, heating mode and basic schematic diagram [1].

Device Form	Heating Method (Temperature Range, °C)	Principle Diagram
Tubular furnace type	Resistance heating mode (~1000)	
Vertical type	Plate heating mode (~500) Induction heating mode (~1200)	
Cylinder type	Induction heating mode (~1200) Infrared radiation heating mode (~1200)	
Tandem surround type	Plate heating mode (~500) Infrared radiation heating mode (~1200)	

[1] Referrence [45].

Generally speaking, the reaction process of CVD mainly includes these parts: vapor-based reagents generation process, reactant transport process, chemical reaction process and reaction by-product removal process, as shown in Figure 2. Controllable variables of CVD method include gas flow rate and composition, deposition temperature, pressure, vacuum chamber shape, deposition time, and substrate material. In particular, basis of the relation between the gas diffusion constant, growth temperature, and gas pressure, the gas diffusion rate can be effectively tuned by an optimal combination of temperature and pressure. As such, CVD has the advantages of precise composition control, complete crystal deposition of thin films, large size and multi-substrate deposition, and deposition on complex substrates [40–45].

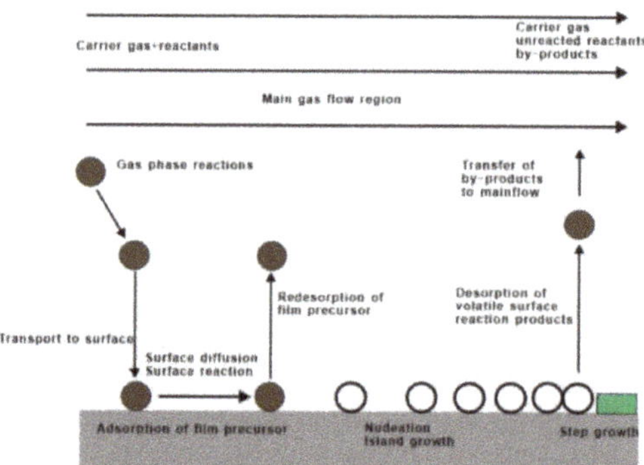

Figure 2. Basic reaction diagram of chemical vapor deposition.

3. Synthesis and Application of PPMs Using the CVD Methods

In the past 10 years, the perovskite photovoltaic film and low dimensional materials and their constructed devices have become a hot topic. Great progress has been made in the preparation and optimization of materials and devices. However, there are still having many problems (such as poor stability, poor repeatability, toxicity, and so on) and challenges to solve. In this part, we aim to clarify why CVD technology can improve the properties of perovskite photovoltaic thin film materials, especially the stability and structure design of the materials, so as to optimize the performance of related devices.

3.1. Perovskite Photovoltaic Film Synthesis through a Variety of CVD Technologies

3.1.1. Atmospheric Pressure and Low Pressure CVD (APCVD and LPCVD) Method

APCVD is a promising approach for producing high quality and large area film materials. In the last few years, one or two step APCVD methods have been used to fabricate perovskite films [46–52]. In 2015, one-step CVD method was used to fabricate planar heterojunction PSCs. By systematically optimizing the CVD parameters, such as temperature and growth time, high quality perovskite films of $CH_3NH_3PbI_3$ and $CH_3NH_3PbI_{3-x}Cl_x$ were obtained. The solar power conversion efficiency (PCE) can be up to 11.1% [46]. At the same year, a two-step sequential tubular CVD (STCVD) method was first employed to fabricate $MAPbI_3$ (MA=CH_3NH_3) material under open-air conditions and PSCs [47]. The measurement results showed that the device performance via STCVD method is significantly enhanced comparing with that of by the other methods of synthesis. For instance, the device Voc can be boost from 0.915 V to 1.001 V, with an impressive enhancement of 86 mV. The best PCE is greatly improved from 10.29% to 13.76%. That is to say, the STCVD method shows high potential to be applied in the commercial production of PSCs, due to the characteristics of high efficiency, stable feature, and low cost. Since then, atmospheric CVD technology has been used to prepare various perovskite films, such as $CsFAPbI_3$, $CsFAPbIBr$, $HC(NH_2)_2PbI_{3x}Cl_x$, and so on [48–52]. These reported works prove that APCVD is a promising technique because it uses a scalable vapor based growth process and the resultant modules maintain a high steady state power at larger areas when compared to modules grown by a reference solution process [22–24].

LPCVD are important thin film deposition techniques based on adsorption and subsequent surface reactions of precursor molecules in a high vacuum environment [53]. The schematic diagram of LPCVD instrument is shown in Figure 3. LPCVD is first introduced into fabrication of $MAPbI_3$ perovskites by Luo group in 2015 [34]. Experimental results show that the prepared $MAPbI_3$ films under low pressure have good crystallization, strong absorption, high stability, and long carrier diffusion length. Uniform and well-defined $MAPbI_3$ layers are observed to be fully covered on the substrates, with a roughness of 19.6 nm and grain size up to 500 nm. Next year, Chen et al. have developed a different one-zone LPCVD to fabricate perovskites [54]. In this work, MAI and PbI_2 are both loaded on a capped graphite boat, and then reacted at 120 °C for 60 or 120 min under a pressure of 133.3 Pa. In this way, efficient PSCs of perovskite modules with a PCE of 6.22% on an active area of 8.4 cm^2 are obtained. In addition, the crucial dependence of working pressure on the film formation is also revealed. Then, Qi group synthesize the $FAPbI_3$ using the CVD with a pressure of ~100 Pa and a flow of drying nitrogen [55,56]. Cui group demonstrated an organic cation exchange concept for the preparation of high-quality a-$FAPbI_3$ films using the CVD method under 10^{-2} Pa [57]. The PCE of the device was 12.4% based on the mesoscopic structure with no hysteresis. Since then, various PMs such as $CH_3NH_3PbI_3$, $(CH_3NH3)_3Bi_2I_9$, and $CsPbBr_3$ have been successfully prepared in low-pressure or ultra-low-pressure CVD systems [58–67]. The published works in recent years using the LPCVD method to fabricated perovskite materials are summarized and listed in Table 2.

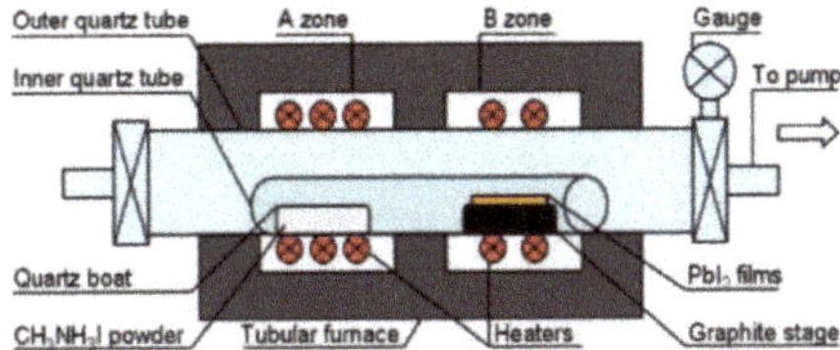

Figure 3. Schematic diagram of LPCVD instrument. Reproduced with permission from reference [34].

Table 2. Perovskite materials were synthesized by LPCVD technique.

Perovskite Material	Pressure	PCE (%)	Ref.
$CH_3NH_3PbI_3$	1 Torr	14.99 (Mesoscopic) 15.37 (Planar)	[54]
FAI (Formamidinium iodide)	2×10^{-2} Pa	14.2	[55]
$CH_3NH_3PbI_3$	1×10^{-3} Pa	15.6	[56]
α-$FAPbI_3$	10^{-2} Pa	12.4	[57]
AMX_3	10^{-5}–10^{-6} mbar	~	[58]
$CH_3NH_3PbBrI_{3-x}$	170 Torr	~	[59]
$CH_3NH_3I(MAI)$	~	16.42	[60]
$CH_3NH_3PbI_3$	10 hPa	~	[61]
$CsPbX_3$ (X = Cl, Br, I)	4.8(4.8,5.2) Torr	5.9 (10.0, 8.3)	[62]
$(CH_3NH_3)_3Bi_2I_9$	10^{-6} hPa	0.047	[63]
$CH_3NH_3PbI_3$	10^4 Pa	~	[64]
$CH_3NH_3PbI_3$	1 Torr	7.9	[65]
$CsPbBr_3$	150 Pa	~	[66]
$(CH_3NH_3)_3Bi_2I_9$	10 hpa	0.047	[67]

3.1.2. Aerosol Assisted CVD (AACVD) Method

Compared with other deposition methods, AACVD is a low-cost and scalable technique. AACVD process occurs under ambient pressure and requires moderately volatile precursors, which should be a very attractive film fabrication technology and has a good potential for scale-up. Lewis et al. first use the AACVD method to the fabrication of MAPbBr$_3$ perovskite film in 2014 [68]. The dilute nebulized precursors are transported by N$_2$ gas into the tubular reaction furnace. The yellow perovskite film is deposited at 250 °C. Subsequently, phase pure, compositionally uniform MAPbI$_3$ films on large glass substrate are prepared by Palgrave in 2015 using the AACVD method [69]. Then, several research groups have developed a one-step AACVD method to deposit perovskite thin films [70,71]. For instance, Liu et al. present a novel NH$_4$Cl + CH$_3$NH$_3$I mixing vapor-assisted CVD method to realize the low-temperature and rapid preparation of perovskite layers [71].

Next year, Binions et al. develop a two-step sequential AACVD method to deposit MAPbI$_3$ film [72]. To obtain uniform and thick enough perovskite film, a modified three-step AACVD method was proposed by Afzal et al. [73]. In the three-step AACVD deposition technique, a cold-walled horizontal-bed AACVD reactor was used to deposit perovskite MAPbI$_3$ film. To overcome the solubility limitation of bromide in conventional polar solvent, a novel Br$_2$-vapor-assisted CVD method was elaborately design to realize the fast anion-exchange from CsPbI$_3$ to CsPbBr$_3$, and thus structural stability of inorganic perovskites is enhanced accordingly [74]. PSCs based on these perovskite materials demonstrate good long term stability and give ~90% of initial efficiencies even after 21 days of exposure to air.

3.1.3. Hybrid Physical CVD (HPCVD) Method

In 2014, Qi et al. first introduced HPCVD technology for the effective deposition of a perovskite layer [40]. In this method, PbX$_2$ was evaporated in a high-vacuum environment to obtain uniform films, and methyl ammonium iodide (MAI) was then heated to 185 °C and transported into the reaction site by N$_2$ gas to fabricate perovskite solar cells. Next year, high-quality CH$_3$NH$_3$PbI$_3$ films and corresponding solar cells were prepared by HPCVD method with a reaction temperature of 100 °C [75]. By optimizing the reaction temperature and pressure, efficient CH$_3$NH$_3$PbI$_3$ solar cells were fabricated with high conversion efficiency up to 12.3%. In the same year, Peng etc. [76] also synthesized the high-quality CH$_3$NH$_3$PbI$_3$ films using the HPCVD process in a vacuum and isothermal environment. CH$_3$NH$_3$PbI$_3$ solar cells with high PCE up to 14.7% were fabricated at 82 °C. In 2016, all-vacuum processed methyl ammonium lead halide perovskite by a sequence of physical vapor deposition of PbI$_2$ and CVD of CH$_3$NH$_3$I under a static atmosphere was fabricated [77,78]. A dependence of residual PbI$_2$ on the solar cells performance is displayed, while photovoltaic devices with efficiency up to 11.6% were achieved. It should be pointed out that, the diffusion rate can be varied over a wide range by controlling its gas pressure when the growth temperature is limited to reasonable value for the HPCVD system. Thus, high-quality perovskite films with fully covered, smooth surfaces are achieved by using the HPCVD method [37,79]. A hybrid CVD with cation exchange method is developed for preparation of Cs-substituted mixed cation perovskite films [80–82]. This technique shows a high potential toward scaling-up the Cs-substituted perovskite from lab solar cell scale 1.5 cm^2 to module scale 5 cm^2 with a high module PCE of 14.6% (12.0 cm^2 active areas). Cs$_{0.07}$FA$_{0.93}$PbI$_3$ PSC prepared by this method shows 14.0% PCE after 1200 min steady-state measurement, demonstrating the promising device stability achieved by this perovskite fabrication technique [83].

3.2. Low Dimensional Perovskite Photovoltaic Materials Synthesized by CVD Technologies

Over the last five years, there has been tremendous progress in the development of nanoscale PPMs which possess a wide range of band gaps and tunable optical and electronic properties. These excellent photoelectric properties and ultra-high density of nanostructured PPMs can serve as ideal candidates for [81–89] solar cells.

The emerging 2D PPMs are attracting more interest due to the long charge carrier lifetime, high photoluminescence quantum efficiency, and great defect tolerance [90]. Recently, Ha et al. first developed CVD synthesis of perovskite nanoplatelets on mica substrate [20]. This method was further employed by Su group [91] to push the ample thickness (below 10 nm) on mica substrates.

The weak material–substrate interaction and low cohesive energy of the perovskite lead to the growth of large-scale ultrathin 2D crystals. Wang et al. [92] used a dual precursor CVD method to grown $MAPbCl_3$ 2D sheet with a thickness of 8.7 nm and lateral dimension of over 20 mm. Nanosheets with thicknesses as low as 1.3 nm have been synthesized using CVD at a low pressure of 40 Torr and 120 °C. Bao and other research teams [21,93–95] use the CVD to create 2D $CH_3NH_3PbI_3$ perovskites. Lan et al. show that process pressure can be used to grow PbI_2 with high crystallinity on roughened surfaces [96]. CVD was also used to convert the as-synthesized nanosheets to $CH_3NH_3PbI_3$. Chen et al. synthesized ultrathin $(BA)_2(MA)_{n-1}PbnBr_{3n+1}$ perovskites with thickness down to 4.2 nm and lateral dimension up to 57 mm [97]. These results suggest that the one-step and two-step vapor phase synthetic approaches are powerful in prepare low dimensional PMs.

Particularly, thin 2D materials can be used as hole extraction layers in organolead halide PSCs [98,99]. MoS_2 and WS_2 layers with a polycrystalline structure were synthesized by a thermal CVD system at 950 °C and 1 Torr pressure. The PCE of the MoS_2- and WS_2-based PSCs were 9.53% and 8.02%, respectively. These results suggest that 2D materials such as MoS_2 and WS_2 can be promising candidates for the formation of hole extraction layers in the PSCs. $FAPb(Br_xI_{1-x})_3$ nanoplatelets with gradient bandgap are fabricated using CVD method [100]. During the whole vapor conversion process, CVD furnace temperature was 140 °C, and pressure was about 500 mTorr with a controlled Ar gas flow rate of 35 sccm (standard cubic centimeters per minute) and reaction time 2 h.

4. Electrode, Window, Blocking and Electron Transport Layer Materials of Perovskite Devices Prepared by CVD Technologies

Interfacing perovskites with other 1D or 2D materials including graphene, carbon nanotubes (CNTs), SnO_2, ZnO, CuO_x, and CdS significantly broadens the application range of the perovskite materials. Therefore, using various CVD technologies to prepare high-quality electrode, window, blocking, and electron transport layer materials of perovskite devices can effectively promote device design and enhances the performance of the functional devices.

4.1. APCVD Method

CVD synthesized graphene films can be used as electrodes in normal structure PSCs to replace the traditional conductive fluorine-doped or indium tin oxide (FTO or ITO) electrodes, either on rigid substrates or on flexible substrates [101–115]. Luo et al. [101] prepared flexible PSCs with a PCE of 11.9% using the graphene as bottom electrode and CNTs as top electrode. Meng et al. [102] synthesized the high quality graphene using CVD method as bottom electrodes in normal structure PSC. An efficiency of 13.93% was obtained from the reverse scan of the current density–voltage (J–V) curves. In 2015, Sung et al. [103] first reported the adoption of graphene as bottom electrodes in inverted structure PSCs. In 2016, Liu et al. [104] also successfully fabricated the flexible PSCs with double layer graphene films as bottom electrodes. The reasonable transparency is ~90% in the visible range, which is much higher than that of ITO electrode [105,106]. The semitransparent PSC using the graphene as top electrode is shown in Figure 4. The following year, Yoon and his collaborators [107] reported super-flexible PSCs by using MoO_3-modified single-layer CVD graphene as bottom electrodes with a maximum PCE of 16.8%. The graphene-based flexible PSCs can maintain over 90% of the original efficiencies after bending test of 1000 cycles at a bending radius of 2 mm. In order to attain higher efficiencies of the graphene-based PSCs, how to further reduce the sheet resistance of graphene films is a key problem. Im and his coworkers employed $AuCl_3$-doped graphene [116–118] and amide TFSA-doped graphene [109] as bottom electrodes to make super-flexible PSCs. After highly p-type chemical doping with two dopants, the sheet resistance of single layer graphene film decreased significantly to approximately $100\ \Omega\ sq^{-1}$, which is much lower than the monolayer graphene. Lang et al. [110] implemented large-area CVD graphene films to obtain semitransparent PSCs and achieving a champion PCE of 13.2% in the final tandem cells. Zhou et al. [111] successfully fabricated terminal perovskite/Si tandem PSCs with a high efficiency of 18.1%. In addition, graphene can also be used as electron transport and blocking layer [112].

Single layer graphene was introduced into PSCs as an air and metal blocking layer to protect the perovskite layer, because the graphene is electrically conductive and can block the metal ions, oxygen, and water into the perovskite layer. Therefore, PSCs including a graphene layer showed a significantly enhanced stability under ambient conditions or thermal annealing process [113]. Therefore, we can see it clearly that the conductivity and quality of the transferred graphene films can severely impact the device performance of PSCs.

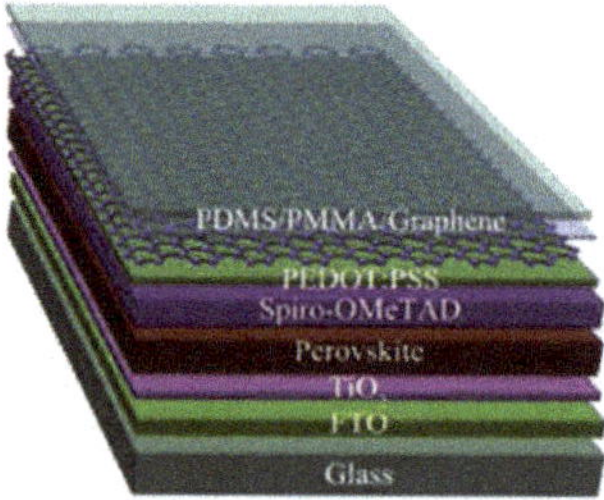

Figure 4. Schematic diagram of the semitransparent PSC using the graphene as top electrode. Reproduced with permission from reference [105].

Except the graphene, CNTS (network films) prepared by CVD method are also used as electrodes of perovskite devices [114,115]. CNT network films were synthesized using the floating catalyst CVD method using a tube furnace. The transparent CNTs top electrode acts as hole collecting layer and light transmission. The PSCs with a PCE up to 8.31% has been achieved. The published works of using the graphene and CNT materials as top and bottom electrodes are summarized and listed in Table 3.

Table 3. Summary of the PSCs with graphene and CNTs electrodes.

Electrode Material	Device Structures	PCE (%)	Ref.
Graphene CNTS	FET/Graphene/TiO$_2$/PCBM/MAPbI$_3$/Spiro-OMeTAD/CNTs	11.9	[101]
Graphene	Quartz/graphene/C$_{60}$/MAPbI$_3$/carbon	13.93	[102]
Graphene	Glass/graphene/MoO$_3$/PEDOT:PSS/MAPbI$_3$/C$_{60}$/BCP/LiF/Al	17.1	[103]
Graphene	PET/ZEOCOAT/graphene/P$_3$HT/MAPbI$_3$/PC71BM/Ag	11.5	[104]
Graphene	FTO/TiO$_2$/MAPbI3-xClx/Spiro-OMeTAD/PEDOT:PSS/graphene	12.37	[105]
Graphene	PET/graphene/PEDOT:PSS/MAPbI$_3$/PCBM/Al	13.94	[106]
Graphene	PEN/graphene/MoO$_3$/PEDOT:PSS/MAPbI$_3$/C$_{60}$/BCP/LiF/Al	16.8	[107]
Graphene	PET/graphene/PEDOT:PSS/FAPbI$_{3-}$xBrx/PCBM/Al	17.9	[108]
Graphene	Glass or PDMS/graphene/PEDOT:PSS/FAPbI$_{3-}$xBrx/PCBM/Al	18.3	[109]
Graphene	FTO/TiO$_2$/MAPbI$_3$/Spiro-OMeTAD/graphene	6.2	[110]
Graphene	FTO/TiO$_2$/MAPbI$_{3-}$xClx/Spiro-OMeTAD/PEDOT:PSS/graphene	11.8	[111]
CNTs	Tifoil/CH$_3$NH$_3$PbI$_3$/TiO$_2$NTs/Spiro-OMeTAD/CNTs	8.31	[114]
CNTs	Glass/FTO/TiO$_2$/CH$_3$NH$_3$PbI$_3$/CNTs	3.88	[115]

4.2. PECVD Method

Large area and functional TiO$_{2-x}$ films using as the hole blocking electron transport layers in PSC architectures are synthesized by roll to roll PECVD system [116]. The PECVD system has a dual laminar flow design to provide two reaction zones to improve net growth rate and uniformity and minimize the entrainment of surrounding air. Each reaction zone incorporates a dielectric barrier system to provide plasma activation and extraction of waste products to avoid contamination. The film thickness can be effectively controlled by the number of passes under the coating head. The experimental results showed that PECVD is capable for producing 50 nm TiO$_{2-x}$ layers with sufficient quality and uniformity for use in perovskite based photovoltaic devices and increasing cell efficiency further.

4.3. Pulsed CVD Method

Pulsed CVD was first used by Bush et al. to prepare SnO_2/ZTO window layer of perovskite/silicon tandem solar cells devices [117]. The vapor processes can produce a compact, uniform, and highly transparent SnO_2/ZTO bilayer with efficient hole-blocking ability and sputter buffer layer properties. The devices have a good thermal and ambient stability compared with the standard encapsulation such as glass. The perovskite cells were coupled with silicon heterojunction bottom cells and ITO layers. The resulting tandem reached an efficiency of 23.6% with no hysteresis. The maximum power can be stabilized over more than 30 minutes under illumination.

Then, in order to solve the problems of high thermal mismatch and low fracture level of PSCs and improve the conversion efficiency of the devices, Cheacharoen et al. prepared the 4 nm SnO_2 and 2 nm Zinc oxide tin layer by pulsed CVD method at 100 °C [118]. The stabilities of the encapsulated PSCs were measured by the International Electrotechnical Commission (IEC) 61646 standard test of 200 temperature cycles from −40 °C to 85 °C and observed no visible delamination and less than a 10% change in performance.

4.4. Physical Chemical Vapor Deposition (PCVD) Method

CdS nanorods arrays and thin film as an electron transport layer in PSCs are fabricated by PCVD method [119,120]. A parametric optimization of the CdS layer thickness and an advantageous in situ Cd-doping of the perovskite thin film lead to a high PCE of ~14.68%. In particular, this lower synthesis temperature (<85 °C) is convenient for implementation a flexible PSCs directly upon plastic PET substrate. Then, 3D architecture PSCs using the CdS nanorods arrays as an electron transport layer were designed and prepared via a layer-by-layer PCVD process. The CdS NRs not only provided a scaffold to the perovskite film, but also increased the interfacial contact between the perovskite film and electron transport layer. As an optimized result, a high PEC of 12.46% with a short-circuit current density of 19.88 mA/cm^{-2}, an open-circuit voltage of 1.01 V was obtained.

4.5. AACVD Method

As mentioned earlier, CNT film has shown to be very effective in replacing metal electrodes and enhancing the stability of PSCs in air environment. Randomly oriented CNT networks with high purity and long nanotube bundle length are synthesized by the AACVD method [121]. Then the carbon-sandwiched perovskite device was studied by using the CVD synthesized CNT on top of MAPbI$_3$. Characterization results show that the encapsulated device (with a structure ITO/C$_{60}$/MAPbI$_3$/CNT) shows high stability against both air and light, with around 90% of the initial efficiency after 2000 h under actual operation conditions.

As a variant of conventional CVD process, AACVD allows for the use of non-volatile metal precursors as the limiting factor is soluble substance rather than volatile substance. Furthermore, the AACVD growth rate could reduce to one comparable or lower than that of atomic layer deposition, making it attractive for ultra-thin film growth. In 2018, ultrathin and compact ZnO films have been successfully deposited on FTO electrodes via AACVD method [122]. ZnO films were carried out in a horizontal bed cold-walled tubular reactor with a size 170 × 60 mm. The deposition temperature was set at 350 °C and the reaction time was 8 mins. After deposition, samples were annealed in a tube furnace at 500 °C for 1 hour under argon atmosphere. Planar PSCs with conventional material $CH_3NH_3PbI_3$ were fabricated under ambient conditions and best PCE of 11.75% was achieved. Also last year, ultra-thin CuOx coatings using as a hole extraction layer in inverted PSCs was prepared by AACVD technique [123]. The resulting $CH_3NH_3PbI_3$ PSC achieves a maximum PCE of 8.26%, demonstrating the efficient hole transporting ability in the CuO_x coatings.

5. Conclusions and Future Outlook

In summary, we have made a review on the preparation and application of perovskite photovoltaic and related device materials by a variety of CVD technologies—including APCVD, LPCVD, HPCVD, HCVD, AACVD, PECVD, and pulsed CVD—and the most recent progress was highlighted over the past five years. Perovskite photovoltaic materials and related device properties with different compositions, morphologies, structures, and dimensions are all discussed. Obviously, it is observed that high stability, high controllability, and high scalability PPMS and high performance devices could be achieved by CVD method through controlling gas flow rate, growth temperature, pressure, and reaction time of the reaction systems. For example, high efficient PSCs with a PCE 23.6% have been achieved by utilizing CVD method to prepare needed materials. Thus, CVD methods owning high stability, low-cost, and scalable features are one of the best choices to prepare high quality perovskite photovoltaic materials, which could be applied for potential industry applications of building PSCs with high efficiency.

However, just as impressive as the progress has been, there are novel opportunities and challenges remaining:

1. Systematic study of synthesis mechanism of CVD method

There is no doubt that great progress has been made in the preparation and optimization of PPMS and devices using various CVD technologies. However, scientists are mainly concerned about how to realize materials and devices with CVD method, and lack of attention and in-depth research on the physical mechanism of synthesis. Understanding the mechanism behind the formation of these perovskite photovoltaic materials will help researchers to come up with effective strategies to combat the emerging challenges of this family of materials, such as stability under ambient conditions and toxicity, towards next generation applications in photovoltaics and optoelectronics.

2. Stabilization

A major limiting factor for the commercialization of perovskite photovoltaic devices is the low stability of the materials. They react very easily with water in the air and are easily oxidized by oxygen in the air. At the same time, the thermal stability of the material is poor, easy to decompose when heated. The longest reported stabilization time is now in the hundreds of days. The stability problems will be amplified at the nanoscale due to the relative surface area being much larger compared to the bulk materials. Thus, stabilizing of the materials and devices is critically important for the perovskites. Selecting ideal encapsulation and protection materials for the perovskites during the process of fabrication is important.

3. Novel perovskite structures

As we mentioned in the introduction, the perovskite structure takes the form of ABX_3. Within this framework, lead and tin are the most commonly used in B position, resulting in their physical properties being unable to be tuned to a precise extent. Searching for novel complex perovskite structures may solve the problem. In particular, many groups have demonstrated that the optical and electrical properties of perovskite photovoltaic materials greatly depend on their dimensions. Thus, ability to fully manipulate their dimensions will be vital for understanding the structure–property–device behavior relationship.

4. Interfaces and heterojunctions

Low-dimensional perovskite materials possess large and well-defined surfaces, and such surfaces can interact strongly with another material's surface when bringing these two materials together to form a heterojunction. Perovskite photovoltaic materials interfacing with low-dimensional graphene, CNTs, SnO_2, ZnO, CuO_x, and CdS materials using the CVD method have been proved can significantly broadens the application and enhances the performance of the functional devices. However, directly growing a lateral heterojunction along the edge of the perovskite low-dimensional materials remains challenging. More complicated heterojunctions with better controlled structures, compositions, and performance are desired in the near future.

CVD technology has shown some advantages and potential applications in preparing perovskite photovoltaic materials and devices than other synthetic methods. It could benefit from the controllable and optimizable growth temperature, pressure, and the moderate gas-phase reaction rate in CVD process. Therefore, the continued development of cost-effective CVD technology is crucial to further commercialization of perovskite photovoltaic devices. This method must draw deep attention from a wide range of researchers and we also hope that it could be vigorously promoted in perovskite materials and devices.

Author Contributions: Conceptualization: L.C., Z.G. and W.G.; writing—original draft preparation: X.L. and Y.L.; writing—review and editing: Z.G. and L.Z.; project administration: L.C., Z.G. and L.Z.

Funding: This research was funded by Natural Science Foundation of China (nos. 11404246, 51675517, and 61874133), STS Projects of the Chinese Academy of Sciences (no. KFJ-STS-SCYD-217), Natural Science Foundation of Shandong Province (no. ZR2018LA014 and ZR2019QEE038), Key Research and Development Plan of Shandong Province (2019GGX101073) and Higher School Science and Technology Plan of Shandong Province (J17KA188), Key Research and Development Program of Jiangsu Province (no. BE2018080), and Tianjin Science and Technology Project (no. 19YFYSQY00040).

Acknowledgments: The authors thank Wang cong and Chen Disheng for their meaningful discussion.

Conflicts of Interest: The authors declare no conflict of interest.

References

1. Rose, G. Beschreibung einiger neuen Mineralien des Urals. *Ann. Phys.* **1839**, *124*, 551–573. [CrossRef]
2. Liu, M.Z.; Johnston, M.B.; Snaith, H.J. Efficient planar heterojunction perovskite solar cells by vapour deposition. *Nature* **2013**, *501*, 395–398. [CrossRef] [PubMed]
3. Lee, M.M.; Teuscher, J.; Miyasaka, T.; Murakami, T.N.; Snaith, H.J. Efficient hybrid solar cells based on meso-superstructured organometal halide perovskites. *Science* **2012**, *338*, 643–647. [CrossRef] [PubMed]
4. Stranks, S.D.; Eperon, G.E.; Grancini, G.; Menelaou, C.; Alcocer, M.J.; Leijtens, T.; Herz, L.M.; Petrozza, A.; Snaith, H.J. Electron-hole diffusion lengths exceeding 1 micrometer in an organometal trihalide perovskite absorber. *Science* **2013**, *342*, 341–344. [CrossRef] [PubMed]
5. Liu, D.; Kelly, T.L. Perovskite solar cells with a planar heterojunction structure prepared using room-temperature solution processing techniques. *Nat. Photonics* **2014**, *8*, 133–138. [CrossRef]
6. Jang, D.M.; Park, K.; Kim, D.H.; Park, J.; Shojaei, F.; Kang, H.S.; Ahn, J.P.; Lee, J.W.; Song, J.K. Reversible halide exchange reaction of organometal trihalide perovskite colloidal nanocrystals for full-range band gap tuning. *Nano Lett.* **2015**, *15*, 5191–5199. [CrossRef]
7. Dong, R.; Fang, Y.; Chae, J.; Dai, J.; Xiao, Z.; Dong, Q.; Yuan, Y.; Centrone, A.; Zeng, X.C.; Huang, J. High-gain and low-driving-voltage photodetectors based on organolead triiodide perovskites. *Adv. Mater.* **2015**, *27*, 1912–1918. [CrossRef]
8. Veldhuis, S.A.; Boix, P.P.; Yantara, N.; Li, M.; Sum, T.C.; Mathews, N.; Mhaisalkar, S.G. Perovskite materials for light-emitting diodes and lasers. *Adv. Mater.* **2016**, *28*, 6804–6834. [CrossRef]
9. Shan, Q.S.; Song, J.Z.; Zou, Y.S.; Li, J.H.; Xu, L.M.; Xue, J.; Dong, Y.H.; Han, B.N.; Chen, J.W.; Zeng, H.B. High performance metal halide perovskite light-emitting diode: From material design to device optimization. *Small* **2017**, *13*. [CrossRef]
10. Ha, S.T.; Su, R.; Xing, J.; Zhang, Q.; Xiong, Q.H. Metal halide perovskite nanomaterials: Synthesis and applications. *Chem. Sci.* **2017**, *8*, 2522–2536. [CrossRef]
11. Makarov, S.; Furasova, A.; Tiguntseva, E.; Hemmetter, A.; Berestennikov, A.; Pushkarev, A.; Zakhidov, A.; Kivshar, Y. Halide-perovskite resonant nanophotonics. *Adv. Opt. Mater.* **2019**, *7*. [CrossRef]
12. Dong, Q.; Fang, Y.; Shao, Y.; Mulligan, P.; Qiu, J.; Cao, L.; Huang, J. Electron-hole diffusion lengths > 75 μm in solution-grown $CH_3NH_3PbI_3$ single crystals. *Science* **2015**, *347*, 967–970. [CrossRef] [PubMed]

13. Nie, W.; Tsai, H.; Asadpour, R.; Blancon, J.C.; Neukirch, A.J.; Gupta, G.; Crochet, J.J.; Chhowalla, M.; Tretiak, S.; Alam, M.A.; et al. High-efficiency solution-processed perovskite solar cells with millimeter-scale grains. *Science* **2015**, *347*, 522–525. [CrossRef]

14. Xiao, Z.; Bi, C.; Shao, Y.; Dong, Q.; Wang, Q.; Yuan, Y.; Wang, C.; Gao, Y.; Huang, J. Efficient, high yield perovskite photovoltaic devices grown by interdiffusion of solution-processed precursor stacking layers. *Energy Environ. Sci.* **2014**, *7*, 2619–2623. [CrossRef]

15. Ning, Z.; Gong, X.; Comin, R.; Walters, G.; Fan, F.; Voznyy, O.; Yassitepe, E.; Buin, A.; Hoogland, S.; Sargent, E.H. Quantum-dot-in-perovskite solids. *Nature* **2015**, *523*, 324–341. [CrossRef]

16. Deng, W.; Xu, X.Z.; Zhang, X.J.; Zhang, Y.D.; Jin, X.C.; Wang, L.; Lee, S.T.; Jie, J.S. Organometal halide perovskite quantum dot light-emitting diodes. *Adv. Funct. Mater.* **2016**, *26*, 4797–4802. [CrossRef]

17. Horvath, E.; Spina, M.; Szekrenyes, Z.; Kamaras, K.; Gaal, R.; Gachet, D.; Forro, L. Nanowires of methylammonium lead iodide ($CH_3NH_3PbI_3$) prepared by low temperature solution-mediated crystallization. *Nano Lett.* **2014**, *14*, 6761–6766. [CrossRef]

18. Petrov, A.A.; Pellet, N.; Seo, J.Y.; Belich, N.A.; Kovalev, D.Y.; Shevelkov, A.V.; Goodilin, E.A.; Zakeeruddin, S.M.; Tarasov, A.B.; Graetzel, M. New insight into the formation of hybrid perovskite nanowires via structure directing adducts. *Chem. Mater.* **2017**, *29*, 587–594. [CrossRef]

19. Liu, P.; He, X.; Ren, J.; Liao, Q.; Yao, J.; Fu, H. Organic–Inorganic hybrid perovskite nanowire laser arrays. *ACS Nano* **2017**, *11*, 5766–5773. [CrossRef]

20. Ha, S.T.; Liu, X.F.; Zhang, Q.; Giovanni, D.; Sum, T.C.; Xiong, Q.H. Synthesis of organic–Inorganic lead halide perovskite nanoplatelets: Towards high-performance perovskite solar cells and optoelectronic devices. *Adv. Opt. Mater.* **2014**, *2*, 838–844. [CrossRef]

21. Liu, J.Y.; Xue, Y.Z.; Wang, Z.Y.; Xu, Z.Q.; Zheng, C.X.; Weber, B.; Song, J.C.; Wang, Y.S.; Lu, Y.R.; Zhang, Y.P.; et al. Two-dimensional $CH_3NH_3PbI_3$ perovskite: Synthesis and optoelectronic application. *ACS Nano* **2016**, *10*, 3536–3542. [CrossRef] [PubMed]

22. Burschka, J.; Pellet, N.; Moon, S.J.; Humphry-Baker, R.; Gao, P.; Nazeeruddin, M.K.; Grätzel, M. Sequential deposition as a route to high-performance perovskite-sensitized solar cells. *Nature* **2013**, *499*, 316–319. [CrossRef] [PubMed]

23. Zhou, H.P.; Chen, Q.; Li, G.; Luo, S.; Song, T.B.; Duan, H.S.; Hong, Z.R.; You, J.B.; Liu, Y.S.; Yang, Y. Interface engineering of highly efficient perovskite solar cells. *Science* **2014**, *345*, 542–546. [CrossRef] [PubMed]

24. Eperon, G.E.; Burlakov, V.M.; Docampo, P.; Goriely, A.; Snaith, H.J. Morphological control for high performance, solution-processed planar heterojunction perovskite solar cells. *Adv. Funct. Mater.* **2014**, *24*, 151–157. [CrossRef]

25. Dualeh, A.; Tetreault, N.; Moehl, T.; Gao, P.; Nazeeruddin, M.K.; Grätzel, M. Effect of annealing temperature on film morphology of organic–inorganic hybrid perovskite solid-state solar cells. *Adv. Funct. Mater.* **2014**, *24*, 3250–3258. [CrossRef]

26. Xiao, M.; Huang, F.; Huang, W.; Dkhissi, Y.; Zhu, Y.; Etheridge, J.; Weale, A.G.; Bach, U.; Cheng, Y.B.; Spiccia, L. A fast deposition-crystallization procedure for highly efficient lead iodide perovskite thin-film solar cells. *Angew. Chem. Int. Ed.* **2014**, *53*, 9898–9903. [CrossRef] [PubMed]

27. Chen, C.W.; Kang, H.W.; Hsiao, S.Y.; Yang, P.F.; Chiang, K.M.; Lin, H.W. Efficient and uniform planar-type perovskite solar cells by simple sequential vacuum deposition. *Adv. Mater.* **2014**, *246*, 6647–6652. [CrossRef] [PubMed]

28. Longo, G.; Gil-Escrig, L.; Degen, M.J.; Sessolo, M.; Bolink, H.J. Perovskite solar cells prepared by flash evaporation. *Chem. Commun.* **2015**, *51*, 7376–7378. [CrossRef] [PubMed]

29. Deng, Y.; Peng, E.; Shao, Y.; Xiao, Z.; Dong, Q.; Huang, J. Scalable fabrication of efficient organolead trihalide perovskite solar cells with doctor-bladed active layers. *Energy Environ. Sci.* **2015**, *8*, 1544–1550. [CrossRef]

30. Hwang, K.; Jung, Y.S.; Heo, Y.J.; Scholes, F.H.; Watkins, S.E.; Subbiah, J.; Jones, D.J.; Kim, D.Y.; Vak, D. Toward large scale roll-to-roll production of fully printed perovskite solar cells. *Adv. Mater.* **2015**, *27*, 1241–1247. [CrossRef]

31. Barrow, A.T.; Pearson, A.J.; Kwak, C.; Dunbar, A.D.F.; Buckley, A.R.; Lidzey, D.G. Efficient planar heterojunction mixed-halide perovskite solar cells deposited via spray-deposition. *Energy Environ. Sci.* **2014**, *7*, 2944–2950. [CrossRef]

32. Wei, Z.; Chen, H.; Yan, K.; Yang, S. Inkjet printing and instant chemical transformation of a $CH_3NH_3PbI_3$/nanocarbon electrode and interface for planar perovskite solar cells. *Angew. Chem. Int. Ed.* **2014**, *53*, 13239–13243. [CrossRef] [PubMed]

33. Chen, Q.; Zhou, H.; Hong, Z.; Luo, S.; Duan, H.S.; Wang, H.H.; Liu, Y.; Li, G.; Yang, Y. Planar heterojunction perovskite solar cells via vapor-assisted solution process. *J. Am. Chem. Soc.* **2014**, *136*, 622–655. [CrossRef] [PubMed]

34. Luo, P.F.; Liu, Z.F.; Xia, W.; Yuan, C.C.; Cheng, J.G.; Lu, Y.W. Uniform, stable, and efficient planar-heterojunction perovskite solar cells by facile low-pressure chemical vapor deposition under fully open-air conditions. *ACS Appl. Mater. Interfaces* **2015**, *7*, 2708–2714. [CrossRef] [PubMed]

35. Chen, J.; Zhou, S.; Jin, S.; Li, H.; Zhai, T. Crystal organometal halide perovskites with promising optoelectronic applications. *J. Mater. Chem. C* **2016**, *4*, 11–27. [CrossRef]

36. Ono, L.K.; Leyden, M.R.; Wang, S.; Qi, Y. Organometal halide perovskite thin films and solar cells by vapor deposition. *J. Mater. Chem. A* **2016**, *4*, 6693–6713. [CrossRef]

37. Luo, P.F.; Zhou, S.W.; Xia, W.; Cheng, J.G.; Xu, C.X.; Lu, Y.W. Chemical vapor deposition of perovskites for photovoltaic application. *Adv. Mater. Interfaces* **2017**, *4*. [CrossRef]

38. Xu, X.; Wang, W.; Zhou, W.; Shao, Z. Recent advances in novel nanostructuring methods of perovskite electrocatalysts for energy-related applications. *Small Methods* **2018**, *2*. [CrossRef]

39. Jamal, M.S.; Bashar, M.S.; Hasan, A.M.; Almutairi, Z.A.; Alharbi, H.F.; Alharthi, N.H.; Karim, M.R.; Misran, H.; Amin, N.; Sopian, K.B.; et al. Fabrication techniques and morphological analysis of perovskite absorber layer for high-efficiency perovskite solar cell: A review. *Renew. Sustain. Energy Rev.* **2018**, *98*, 469–488. [CrossRef]

40. Leyden, M.R.; Ono, L.K.; Raga, S.R.; Kato, Y.; Wang, S.; Qi, Y. High performance perovskite solar cells by hybrid chemical vapor deposition. *J. Mater. Chem. A* **2014**, *2*, 18742–18745. [CrossRef]

41. Ting, J.M.; Huang, N.Z. Thickening of chemical vapor deposited carbon fiber. *Carbon* **2001**, *39*, 835–839. [CrossRef]

42. Darr, J.A.; Guo, Z.X.; Raman, V.; Bououdina, M.; Rehman, I.U. Metal organic chemical vapour deposition (MOCVD) of bone mineral like carbonated hydroxyapatite coatings. *Chem. Commun.* **2004**, *6*, 696–697. [CrossRef] [PubMed]

43. Li, Y.; Mann, D.; Rolandi, M.; Kim, W.; Ural, A.; Hung, S.; Javey, A.; Cao, J.; Wang, D.; Yenilmez, E.; et al. Preferential growth of semiconducting single-walled carbon nanotubes by a plasma enhanced CVD method. *Nano Lett.* **2004**, *4*, 317–321. [CrossRef]

44. Suresh, A.; Anastasio, D.; Burkey, D.D. Potential of hexyl acrylate monomer as an initiator in photo-initiated CVD. *Chem. Vapor Depos.* **2014**, *20*, 5–7. [CrossRef]

45. Jones, A.C.; Hitchman, M.L. *Chemical Vapour Deposition: Precursors, Processes and Applications*; RSC Publishing: London, UK, 2009.

46. Tavakoli, M.M.; Gu, L.L.; Gao, Y.; Reckmeier, C.; He, J.; Rogach, A.L.; Yao, Y.; Fan, Z.Y. Fabrication of efficient planar perovskite solar cells using a one-step chemical vapor deposition method. *Sci. Rep.* **2015**, *5*, 14083. [CrossRef] [PubMed]

47. Luo, P.F.; Liu, Z.F.; Xia, W.; Yuan, C.C.; Cheng, J.Q.; Xu, C.X.; Lu, Y.W. Chlorine-conducted defect repairmen and seed crystal-mediated vapor growth process for controllable preparation of efficient and stable perovskite solar cells. *J. Mater. Chem. A* **2015**, *3*, 22949–22959. [CrossRef]

48. Leyden, M.R.; Meng, L.Q.; Jiang, Y.; Ono, L.K.; Qiu, L.B.; Juarez-Perez, E.J.; Qin, C.J.; Adachi, C.H.; Qi, Y.B. Methylammonium lead bromide perovskite light-emitting diodes by chemical vapor deposition. *J. Phys. Chem. Lett.* **2017**, *8*, 3193–3198. [CrossRef] [PubMed]

49. Wei, Q.; Bi, H.; Yan, S.; Wang, S.W. Morphology and interface engineering for organic metal halide perovskite-based photovoltaic cells. *Adv. Mater. Interfaces* **2018**, *5*. [CrossRef]

50. Sun, J.C.; Wu, J.; Tong, X.; Wang, Y.N.; Wang, Z.M. Organic/inorganic metal halide perovskite optoelectronic devices beyond solar cells. *Adv. Sci.* **2018**, *5*. [CrossRef]

51. Tavakoli, M.M.; Zakeeruddin, S.M.; Grätzel, M.; Fan, Z.Y. Large-grain tin-rich perovskite films for efficient solar cells via metal alloying technique. *Adv. Mater.* **2018**, *30*. [CrossRef]

52. Swartwout, R.; Hoerantner, M.T.; Bulović, V. Scalable deposition methods for large-area production of perovskite thin films. *Energy Environ. Mater.* **2019**, *2*, 119–143. [CrossRef]

53. Park, N.G. Crystal growth engineering for high efficiency perovskite solar cells. *CrystEngComm* **2016**, *18*, 5977–5985. [CrossRef]

54. Shen, P.S.; Chen, J.S.; Chiang, Y.H.; Li, M.H.; Guo, T.F.; Chen, P. Low-pressure hybrid chemical vapor growth for efficient perovskite solar cells and large-area module. *Adv. Mater. Interfaces* **2016**, *3*. [CrossRef]

55. Leyden, M.R.; Lee, M.V.; Raga, S.R.; Qi, Y.B. Large formamidinium lead trihalide perovskite solar cells using chemical vapor deposition with high reproducibility and tunable chlorine concentrations. *J. Mater. Chem. A* **2015**, *3*, 16097–16103. [CrossRef]

56. Leyden, M.R.; Jiang, Y.; Qi, Y. Chemical vapor deposition grown formamidinium perovskite solar modules with high steady state power and thermal stability. *J. Mater. Chem. A* **2016**, *4*, 13125–13132. [CrossRef]

57. Zhou, Z.M.; Pang, S.P.; Ji, F.X.; Zhang, B.; Cui, G.L. The fabrication of formamidinium lead iodide perovskite thin films via organic cation exchange. *Chem. Commun.* **2016**, *52*, 3828–3831. [CrossRef] [PubMed]

58. Ávila, J.; Momblona, C.; Boix, P.P.; Sessolo, M.; Bolink, H.J. Vapor-deposited perovskites: The route to high-performance solar cell production? *Joule* **2017**, *1*, 431–442. [CrossRef]

59. Wang, Y.P.; Chen, Z.Z.; Deschler, F.; Sun, X.; Lu, T.M.; Wertz, E.A.; Hu, J.M.; Shi, J. Epitaxial halide perovskite lateral double heterostructure. *ACS Nano* **2017**, *11*, 3355–3364. [CrossRef]

60. Tong, G.Q.; Lan, X.Z.; Song, Z.H.; Li, G.P.; Li, H.; Yu, L.W.; Xu, J.; Jiang, Y.; Sheng, Y.; Shi, Y.; et al. Surface-activation modified perovskite crystallization for improving photovoltaic performance. *Mater. Today Energy* **2017**, *5*, 173–180. [CrossRef]

61. Stümmler, D.; Sanders, S.; Pfeiffer, P.; Weingarten, M.; Vescan, A.; Kalisch, H. Direct chemical vapor phase deposition of organometal halide perovskite layers. *MRS Adv.* **2017**, *2*, 1189–1194. [CrossRef]

62. Guo, P.F.; Hossain, M.K.; Shen, X.; Sun, H.B.; Yang, W.C.; Liu, C.P.; Ho, C.Y.; Kwok, C.K.; Tsang, S.W.; Luo, Y.S.; et al. Room-temperature red-green-blue whispering-gallery mode lasing and white-light emission from cesium lead halide perovskite (CsPbX$_3$, X = Cl, Br, I) microstructures. *Adv. Opt. Mater.* **2018**, *6*, 1700993. [CrossRef]

63. Stümmler, D.; Sanders, S.; Pfeiffer, P.; Wickel, N.; Simkus, G.; Heuken, M.; Baumann, P.K.; Vescan, A.; Kalisch, H. Investigation of perovskite solar cells employing chemical vapor deposited methylammonium bismuth iodide layers. *MRS Adv.* **2018**, *3*, 3069–3074. [CrossRef]

64. Lei, Y.; Gu, L.Y.; Yang, X.G.; Zhang, L.L.; Gao, Y.H.; Li, J.X.; Zheng, Z. Fast chemical vapor-solid reaction for synthesizing organometal halide perovskite array thin films for photodetector applications. *J. Alloys Compd.* **2018**, *766*, 933–940. [CrossRef]

65. Tran, V.D.; Pammi, S.V.N.; Dao, V.D.; Choi, H.S.; Yoo, S.G. Chemical vapor deposition in fabrication of robust and highly efficient perovskite solar cells based on single-walled carbon nanotubes counter electrodes. *J. Alloys Compd.* **2018**, *747*, 703–711. [CrossRef]

66. Tian, C.C.; Wang, F.; Wang, Y.P.; Yang, Z.; Chen, X.J.; Mei, J.J.; Liu, H.Z.; Zhao, D.X. Chemical vapor deposition method grown all-inorganic perovskite microcrystals for self-powered photodetectors. *ACS Appl. Mater. Interfaces* **2019**, *11*, 15804–15812. [CrossRef] [PubMed]

67. Stümmler, D.; Sanders, S.; Gerstenberger, F.; Pfeiffer, F.; Simkus, G.; Baumann, P.K.; Heuken, M.; Vescan, M.; Kalisch, H. Reaction engineering of CVD methylammonium bismuth iodide layers for photovoltaic applications. *J. Mater. Res.* **2019**, *34*, 608–615. [CrossRef]

68. Lewis, D.J.; O'Brien, P. Ambient pressure aerosol-assisted chemical vapour deposition of (CH$_3$NH$_3$) PbBr$_3$, an inorganic-organic perovskite important in photovoltaics. *Chem. Commun.* **2014**, *50*, 6319–6321. [CrossRef]

69. Bhachu, D.S.; Scanlon, D.O.; Saban, E.J.; Bronstein, H.; Parkin, I.P.; Carmalt, C.J.; Palgrave, R.G. Scalable route to CH$_3$NH$_3$PbI$_3$ perovskite thin films by aerosol assisted chemical vapour deposition. *J. Mater. Chem. A* **2015**, *3*, 9071–9073. [CrossRef]

70. Afzaal, M.; Yates, H.M. Growth patterns and properties of aerosol-assisted chemical vapor deposition of CH$_3$NH$_3$PbI$_3$ films in a single step. *Surf. Coat. Technol.* **2017**, *321*, 336–340. [CrossRef]

71. Liu, Z.F.; Luo, P.F.; Xia, W.; Zhou, S.W.; Cheng, J.Q.; Sun, L.; Xu, C.X.; Lu, Y.W. Acceleration effect of chlorine in the gas-phase growth process of CH$_3$NH$_3$PbI$_3$(Cl) films for efficient perovskite solar cells. *J. Mater. Chem. C* **2016**, *4*, 6336–6344. [CrossRef]

72. Chen, S.; Briscoe, J.; Shi, Y.; Chen, K.; Wilson, R.M.; Dunn, S.; Binions, R. A simple, low-cost CVD route to high-quality CH$_3$NH$_3$PbI$_3$ perovskite thin films. *CrystEngComm* **2015**, *17*, 7486–7489. [CrossRef]

73. Afzaal, M.; Salhi, B.; Al-Ahmed, A.; Yates, H.; Hakeem, A. Surface-related properties of perovskite CH$_3$NH$_3$PbI$_3$ thin films by aerosol-assisted chemical vapour deposition. *J. Mater. Chem. C* **2017**, *5*, 8366–8370. [CrossRef]

74. Luo, P.F.; Zhou, Y.G.; Zhou, S.W.; Lu, Y.W.; Xu, C.X.; Sun, L. Fast anion-exchange from $CsPbI_3$ to $CsPbBr_3$ via Br_2-vapor-assisted deposition for air-stable all-inorganic perovskite solar cells. *Chem. Eng. J.* **2018**, *343*, 146–154. [CrossRef]

75. Peng, Y.; Jing, G.; Cui, T. High crystalline quality perovskite thin films prepared by a novel hybrid evaporation/CVD technique. *MRS Online Proc. Libr. Arch.* **2015**, *1771*, 187–192. [CrossRef]

76. Peng, Y.; Jing, G.; Cui, T. A hybrid physical-chemical deposition process at ultra-low temperatures for high-performance perovskite solar cells. *J. Mater. Chem. A* **2015**, *3*, 12436–12442. [CrossRef]

77. Ioakeimidis, A.; Christodoulou, C.; Lux-Steiner, M.; Fostiropoulos, K. Effect of PbI_2 deposition rate on two-step PVD/CVD all-vacuum prepared perovskite. *J. Solid State Chem.* **2016**, *244*, 20–24. [CrossRef]

78. Luo, P.F.; Zhou, S.W.; Liu, Z.F.; Xia, W.; Sun, L.; Cheng, J.G.; Xu, C.X.; Lu, Y.W. A novel transformation route from PbS to $CH_3NH_3PbI_3$ for fabricating curved and large-area perovskite films. *Chem. Commun.* **2016**, *52*, 11203–11206. [CrossRef]

79. Hossain, M.I.; Qarony, W.; Jovanov, V.; Tsang, Y.H.; Knipp, D. Nanophotonic design of perovskite/silicon tandem solar cells. *J. Mater. Chem. A* **2018**, *6*, 3625–3633. [CrossRef]

80. Jiang, Y.; Leyden, M.R.; Qiu, L.B.; Wang, S.H.; Ono, L.K.; Wu, Z.F.; Juarez-Perez, E.J.; Qi, Y.B. Combination of hybrid CVD and cation exchange for upscaling Cs-substituted mixed cation perovskite solar cells with high efficiency and stability. *Adv. Funct. Mater.* **2018**, *28*. [CrossRef]

81. Zhang, Y.P.; Liu, J.Y.; Wang, Z.Y.; Xue, Y.Z.; Ou, Q.D.; Polavarapu, L.; Zheng, J.L.; Qi, X.; Bao, Q.L. Synthesis, properties, and optical applications of low-dimensional perovskites. *Chem. Commun.* **2016**, *52*, 13637–13655. [CrossRef]

82. Chen, S.; Shi, G. Two-dimensional materials for halide perovskite-based optoelectronic devices. *Adv. Mater.* **2017**, *29*. [CrossRef] [PubMed]

83. Dou, L. Emerging two-dimensional halide perovskite nanomaterials. *J. Mater. Chem. C* **2017**, *5*, 11165–11173. [CrossRef]

84. Fu, P.F.; Shan, Q.S.; Shang, Y.Q.; Song, J.Z.; Zeng, H.B.; Ning, Z.J.; Gong, J.K. Perovskite nanocrystals: Synthesis, properties and applications. *Sci. Bull.* **2017**, *62*, 369–380. [CrossRef]

85. Yusoff, A.R.B.M.; Nazeeruddin, M.K. Low-dimensional perovskites: From synthesis to stability in perovskite solar cells. *Adv. Energy Mater.* **2018**, *8*. [CrossRef]

86. Hong, K.; Van Le, Q.; Kim, S.Y.; Jang, H.W. Low-dimensional halide perovskites: Review and issues. *J. Mater. Chem. C* **2018**, *6*, 2189–2209. [CrossRef]

87. Huang, K.; Liu, J.; Chang, X.; Fu, L. Integrating properties modification in the synthesis of metal halide perovskites. *Adv. Mater. Technol.* **2019**, *4*. [CrossRef]

88. Lee, K.J.; Turedi, B.; Sinatra, L.; Zhumekenov, A.A.; Maity, P.; Dursun, I.; Naphade, R.; Merdad, N.; Alsalloum, A.; Oh, S.; et al. Perovskite-based artificial multiple quantum wells. *Nano Lett.* **2019**, *19*, 3535–3542. [CrossRef] [PubMed]

89. Lan, C.; Zhou, Z.; Wei, R.; Ho, J.C. Two-dimensional perovskite materials: From synthesis to energy-related applications. *Mater. Today Energy* **2019**, *11*, 61–82. [CrossRef]

90. Huo, C.X.; Cai, B.; Yuan, Z.; Ma, B.W.; Zeng, H.B. Two-dimensional metal halide perovskites: Theory, synthesis, optoelectronics. *Small Methods* **2017**, *1*. [CrossRef]

91. Su, L.; Zhao, Z.X.; Li, H.Y.; Yuan, J.; Wang, Z.L.; Cao, G.Z.; Zhu, G. High-performance organolead halide perovskite-based self-powered triboelectric photodetector. *ACS Nano* **2015**, *9*, 11310–11316. [CrossRef] [PubMed]

92. Wang, Y.P.; Shi, Y.F.; Xin, G.Q.; Lan, J.; Shi, J. Two-dimensional van der Waals epitaxy kinetics in a three-dimensional perovskite halide. *Cryst. Growth Des.* **2015**, *15*, 4741–4749. [CrossRef]

93. Li, P.F.; Shivananju, B.N.; Zhang, Y.P.; Li, S.J.; Bao, Q.L. High performance photodetector based on 2D $CH_3NH_3PbI_3$ perovskite nanosheets. *J. Phys. D Appl. Phys.* **2017**, *50*, 094002. [CrossRef]

94. Qi, X.; Zhang, Y.P.; Ou, Q.D.; Ha, S.T.; Qiu, C.W.; Zhang, H.; Cheng, Y.B.; Xiong, Q.H.; Bao, Q.L. Photonics and optoelectronics of 2D metal-halide perovskites. *Small* **2018**, *14*. [CrossRef] [PubMed]

95. Wen, X.M.; Chen, W.J.; Yang, J.F.; Ou, Q.D.; Yang, T.S.; Zhou, C.H.; Lin, H.; Wang, Z.Y.; Zhang, Y.P.; Conibeer, G.; et al. Role of surface recombination in halide perovskite nanoplatelets. *ACS Appl. Mater. Interfaces* **2018**, *10*, 31586–31593. [CrossRef] [PubMed]

96. Lan, C.; Dong, R.; Zhou, Z.; Shu, L.; Li, D.; Yip, S.; Ho, J.C. Large-scale synthesis of freestanding layer-structured PbI$_2$ and MAPbI$_3$ nanosheets for high-performance photodetection. *Adv. Mater.* **2017**, *29*. [CrossRef] [PubMed]

97. Chen, J.N.; Wang, Y.G.; Gan, L.; He, Y.B.; Li, H.Q.; Zhai, T.Y. Generalized self-doping engineering towards ultrathin and large-sized two-dimensional homologous perovskites. *Angew. Chem. Int. Ed.* **2017**, *56*, 14893–14897. [CrossRef]

98. Kim, Y.G.; Kwon, K.C.; Van Le, Q.; Hong, K.; Jang, H.W.; Kim, S.Y. Atomically thin two-dimensional materials as hole extraction layers in organolead halide perovskite photovoltaic cells. *J. Power Sources* **2016**, *319*, 1–8. [CrossRef]

99. Bai, F.; Qi, J.J.; Li, F.; Fang, Y.Y.; Han, W.P.; Wu, H.L.; Zhang, Y. A high-performance self-powered photodetector based on monolayer MoS$_2$/Perovskite heterostructures. *Adv. Mater. Interfaces* **2018**, *5*. [CrossRef]

100. Zhou, C.H.; Ou, Q.D.; Chen, W.J.; Gan, Z.X.; Wang, J.; Bao, Q.L.; Wen, X.M.; Jia, B.H. Illumination-induced halide segregation in gradient bandgap mixed-halide perovskite nanoplatelets. *Adv. Opt. Mater.* **2018**, *6*. [CrossRef]

101. Luo, Q.; Ma, H.; Hou, Q.Z.; Li, Y.X.; Ren, J.; Dai, X.Z.; Yao, Z.B.; Zhou, Y.; Xiang, L.C.; Du, H.Y.; et al. All-carbon-electrode-based endurable flexible perovskite solar cells. *Adv. Funct. Mater.* **2018**, *28*, 1706777. [CrossRef]

102. Meng, X.Y.; Zhou, J.S.; Hou, J.; Tao, X.; Cheung, S.H.; So, S.K.; Yang, S.H. Versatility of carbon enables all carbon based perovskite solar cells to achieve high efficiency and high stability. *Adv. Mater.* **2018**, *30*. [CrossRef]

103. Sung, H.; Ahn, N.; Jang, M.S.; Lee, J.K.; Yoon, H.; Park, N.G.; Choi, M. Transparent conductive oxide-free graphene-based perovskite solar cells with over 17% efficiency. *Adv. Energy Mater.* **2016**, *6*, 1501873. [CrossRef]

104. Liu, Z.K.; You, P.; Xie, C.; Tang, G.Q.; Yan, F. Ultrathin and flexible perovskite solar cells with graphene transparent electrodes. *Nano Energy* **2016**, *28*, 151–157. [CrossRef]

105. You, P.; Liu, Z.; Tai, Q.D.; Liu, S.H.; Yan, F. Efficient semitransparent perovskite solar cells with graphene electrodes. *Adv. Mater.* **2015**, *27*, 3632–3638. [CrossRef] [PubMed]

106. Kim, S.; Lee, H.S.; Kim, J.M.; Seo, S.W.; Kim, J.H.; Jang, C.W.; Choi, S.H. Effect of layer number on flexible perovskite solar cells employing multiple layers of graphene as transparent conductive electrodes. *J. Alloys Compd.* **2018**, *744*, 404–411. [CrossRef]

107. Yoon, J.; Sung, H.; Lee, G.; Cho, W.; Ahn, N.; Jung, H.S.; Choi, M. Superflexible, high-efficiency perovskite solar cells utilizing graphene electrodes: Towards future foldable power sources. *Energy Environ. Sci.* **2017**, *10*, 337–345. [CrossRef]

108. Heo, J.H.; Shin, D.H.; Jang, M.H.; Lee, M.L.; Kang, M.G.; Im, S.H. Highly flexible, high-performance perovskite solar cells with adhesion promoted AuCl$_3$-doped graphene electrodes. *J. Mater. Chem.* **2017**, *5*, 21146–21152. [CrossRef]

109. Heo, J.H.; Shin, D.H.; Song, D.H.; Kim, D.H.; Lee, S.J.; Im, S.H. Super-flexible bis(trifluoromethanesulfonyl) -amide doped graphene transparent conductive electrodes for photo-stable perovskite solar cells. *J. Mater. Chem.* **2018**, *6*, 8251–8258. [CrossRef]

110. Lang, F.; Gluba, M.A.; Albrecht, S.; Rappich, J.; Korte, L.; Rech, B.; Nickel, N.H. Perovskite solar cells with large-area CVD-graphene for tandem solar cells. *J. Phys. Chem. Lett.* **2015**, *6*, 2745–2750. [CrossRef]

111. Zhou, J.X.; Ren, Z.W.; Li, S.H.; Liang, Z.C.; Surya, C.; Shen, H. Semi-transparent Cl-doped perovskite solar cells with graphene electrodes for tandem application. *Mater. Lett.* **2018**, *220*, 82–85. [CrossRef]

112. You, P.; Tang, G.; Yan, F. Two-dimensional materials in perovskite solar cells. *Mater. Today Energy* **2019**, *11*, 128–158. [CrossRef]

113. Hu, X.H.; Jiang, H.; Li, J.; Ma, J.X.; Yang, D.; Liu, Z.K.; Gao, F.; Liu, S.Z. Air and thermally stable perovskite solar cells with CVD-graphene as the blocking layer. *Nanoscale* **2017**, *9*, 8274–8280. [CrossRef] [PubMed]

114. Wang, X.Y.; Li, Z.; Xu, W.J.; Kulkarni, S.A.; Batabyal, S.K.; Zhang, S.; Cao, A.Y.; Wong, L.H. TiO$_2$ nanotube arrays based flexible perovskite solar cells with transparent carbon nanotube electrode. *Nano Energy* **2015**, *11*, 728–735. [CrossRef]

115. Lee, K.T.; Guo, L.; Park, H. Neutral-and multi-colored semitransparent perovskite solar cells. *Molecules* **2016**, *21*, 475. [CrossRef] [PubMed]

116. Hodgkinson, J.L.; Yates, H.M.; Walter, A.; Sacchetto, D.; Moon, S.J.; Nicolay, S. Roll to roll atmospheric pressure plasma enhanced CVD of titania as a step towards the realisation of large area perovskite solar cell technology. *J. Mater. Chem. C* **2018**, *6*, 1988–1995. [CrossRef]

117. Bush, K.A.; Palmstrom, A.F.; Zhengshan, J.Y.; Boccard, M.; Cheacharoen, R.R.; Mailoa, J.P.; McMeekin, D.P.; Hoye, R.L.Z.; Bailie, C.D.; Leijtens, T.; et al. 23.6%-efficient monolithic perovskite/silicon tandem solar cells with improved stability. *Nat. Energy* **2017**, *2*, 17009. [CrossRef]

118. Cheacharoen, R.R.; Rolston, N.; Harwood, D.; Bush, K.A.; Dauskardt, R.H.; McGehee, M.D. Design and understanding of encapsulated perovskite solar cells to withstand temperature cycling. *Energy Environ. Sci.* **2018**, *11*, 144–150. [CrossRef]

119. Tong, G.Q.; Song, Z.H.; Li, C.D.; Zhao, Y.L.; Yu, L.W.; Xu, J.; Jiang, Y.; Sheng, Y.; Shi, Y.; Chen, K.J. Cadmium-doped flexible perovskite solar cells with a low-cost and low-temperature-processed CdS electron transport layer. *RSC Adv.* **2017**, *7*, 19457–19463. [CrossRef]

120. Song, Z.H.; Tong, G.Q.; Li, H.; Li, G.P.; Ma, S.; Liu, Q.; Jiang, Y. Three-dimensional architecture hybrid perovskite solar cells using CdS nanorod arrays as an electron transport layer. *Nanotechnology* **2017**, *29*, 025401. [CrossRef]

121. Ahn, N.; Jeon, I.; Yoon, J.; Kauppinen, E.I.; Matsuo, Y.; Maruyams, S.; Choi, M. Carbon-sandwiched perovskite solar cell. *J. Mater. Chem. A* **2018**, *6*, 1382–1389. [CrossRef]

122. Chen, S.Q.; Wang, J.S.; Zhang, Z.X.; Briscoe, J.; Warwick, M.E.A.; Li, H.Y.; Hu, P. Aerosol assisted chemical vapour deposition of conformal ZnO compact layers for efficient electron transport in perovskite solar cells. *Mater. Lett.* **2018**, *217*, 251–254. [CrossRef]

123. Zhang, Z.X.; Chen, S.Q.; Li, P.P.; Li, H.Y.; Wu, J.S.; Hu, P.; Wang, J.S. Aerosol-assisted chemical vapor deposition of ultra-thin CuO_x films as hole transport material for planar perovskite solar cells. *Funct. Mater. Lett.* **2018**, *11*, 1850035. [CrossRef]

Review

Advancements, Challenges and Prospects of Chemical Vapour Pressure at Atmospheric Pressure on Vanadium Dioxide Structures

Charalampos Drosos [1] and Dimitra Vernardou [2,3,*]

[1]   Delta Nano-Engineering Solutions Ltd., Paddock Wood, Kent TN12 6EL, UK; info@delta-nano.com
[2]   Center of Materials Technology and Photonics, School of Applied Technology, Technological Educational Institute of Crete, 710 04 Heraklion, Crete, Greece
[3]   Institute of Electronic Structure and Laser, Foundation for Research & Technology-Hellas, P.O. Box 1527, Vassilika Vouton, 711 10 Heraklion, Crete, Greece
*   Correspondence: dimitra@iesl.forth.gr; Tel.: +30-2810-379-774

Received: 1 February 2018; Accepted: 27 February 2018; Published: 5 March 2018

Abstract: Vanadium (IV) oxide (VO_2) layers have received extensive interest for applications in smart windows to batteries and gas sensors due to the multi-phases of the oxide. Among the methods utilized for their growth, chemical vapour deposition is a technology that is proven to be industrially competitive because of its simplicity when performed at atmospheric pressure (APCVD). APCVD's success has shown that it is possible to create tough and stable materials in which their stoichiometry may be precisely controlled. Initially, we give a brief overview of the basic processes taking place during this procedure. Then, we present recent progress on experimental procedures for isolating different polymorphs of VO_2. We outline emerging techniques and processes that yield in optimum characteristics for potentially useful layers. Finally, we discuss the possibility to grow 2D VO_2 by APCVD.

Keywords: APCVD; VO_2; processing parameters; 2D

---

1. Chemical Vapour Deposition

1.1. General Information

CVD is a practical method of atomistic or near atomistic deposition having the ability to synthesize well-controlled dimensions and structures at reasonably low temperatures, high purity and in multiple formats such as single layer, multi-layer, composite and finally functional coatings. In its simplest incarnation, CVD encompasses a single precursor gas flowing into a chamber containing the substrate to be coated. Although, there are exceptions, the vapour of the reactive compound, usually an easily volatilized liquid or in some cases a solid, is sublimed directly and transported to the reaction zone by a carrier gas. A thin film is then deposited by chemical reaction or decomposition of the gas mixture on the substrate surface or in its vicinity at a defined temperature.

The precursors used within a variety of CVD techniques can be single source or dual source in origin. Single source precursors contain all the groups/elements required for successive thin film production. On the other hand, dual source precursors involve the interaction between multiple precursors for the synthesis of thin films. In each case, it is vital for production of thin films to deliver the gas phase precursors with a carrier gas. The most common carrier gases are N_2, He or Ar, especially when highly reactive or pyrophoric reactants are used and in some cases, reactions entail an energy input from the carrier gas, e.g., H_2 or O_2 enrichment.

Reactor systems in CVD processes must allow controlled transport of the reactant and diluent gases to the reaction zone, maintain a defined substrate temperature and safely remove the gaseous by-products. These functions should be fulfilled with sufficient control and maximal effectiveness, which requires optimum engineering design and automation. The reactor in which the thin film deposition actually takes place is the essential part of the system and must be designed according to the specific chemical process parameters. To coat layers using Chemical Vapour Deposition at Atmospheric Pressure (APCVD), four basic types can be classified according to their gas flow and operation principles:

1. Horizontal tube displacement flow type.
2. Rotary vertical batch type.
3. Continuous—deposition type using premixed gas flow.
4. Continuous—deposition type employing separate gas streams.

1.2. CVD Processes

Any CVD process including APCVD involves the subsequent operations. First, the reacting gas is directed into the reactor. The gas moves towards its thermal equilibrium temperature and composition through gas-phase collisions and reactions. Near-equilibrated species are then transported to the reaction surface, the surface chemical reactions commence and the thin film is formed. The processes are summarized below (Figure 1) [1]:

1. Creation of active gaseous reactants.
2. Transport of the precursor to the CVD reactor.
3. Decomposition of gas phase precursor to remove gaseous by-products and grow reactive intermediates.
4. Gaseous reactants transportation onto substrate area.
5. Surface diffusion for nucleation and thin film growth.
6. Desorption of by-products and mass transport away from active reactive zone.

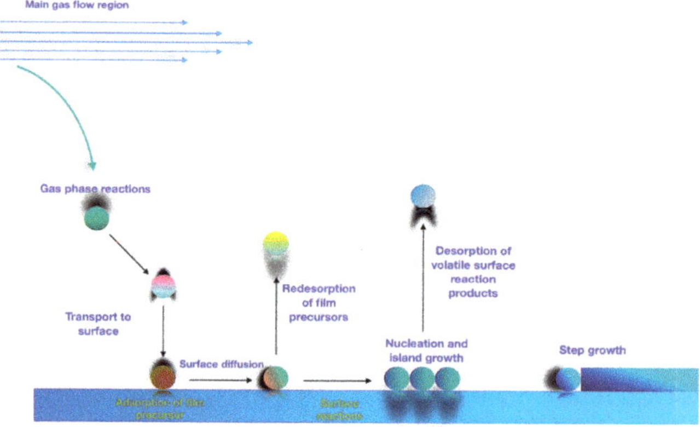

Figure 1. Schematic view of CVD (Chemical Vapour Deposition) process [1].

1.3. APCVD

A schematic presentation of the APCVD system is shown in Figure 2. It is designed with no joints in all outlet lines to avoid blocking. A flow of inert gas, usually nitrogen, is passed through the apparatus during all operations. The amount of the precursor delivered into the reactor is calculated from the Equation (1)

$$a = \frac{VP \times F}{(760 - VP) \times 24.4} \tag{1}$$

where a, is the amount of precursor (mol min^{-1}), VP, is the vapour pressure of precursor at the bubbler's temperature (mm Hg), F, is the nitrogen flow rate through the bubbler (L min^{-1}) and 24.4, is a constant for the molar volume of an ideal gas at standard temperature and pressure (L mol^{-1}).

In a typical APCVD experiment, once all temperatures are stabilized over time, the N_2 is passed through the bubblers and then the precursor gas flow rate is directed into the mixing chamber where the mixture begins in order to be utilized before entering the reaction chamber for the deposition to take place. Once the allotted time is complete, the precursor bubbler is closed. The reactor heater is turned off and the substrate is allowed to cool down under an atmosphere of N_2. Ideally, the carrier gas inlet flows should be fully saturated with precursor vapour; this can be achieved with knowledge of the precursor volatility and vapour pressure and then controlled by the carrier gas flow and bubbler temperature using flow meters and heating jackets.

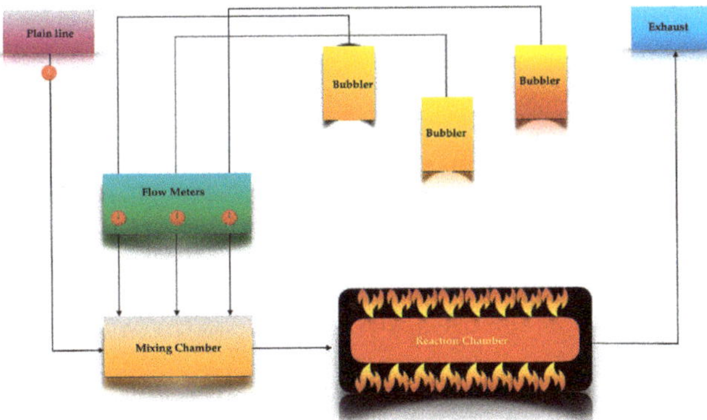

Figure 2. Schematic presentation of an APCVD (Chemical Vapour Deposition at Atmospheric Pressure) system [1].

2. Vanadium Oxides

The binary Vanadium-Oxygen phase diagram consists of a large number of phases between V_2O_3 and VO_2 of the form V_nO_{2n-1} commonly known as the Magneli phases [2] that exhibit distinctive electrical and optical properties. The variety of Vanadium-Oxygen stoichiometries emerges from the ability of vanadium atoms to adopt multiple oxidation states, which consequently results in synthetic challenges to control the structure of the materials [3].

More than ten kinds of crystalline phases of VO_2 have been reported elsewhere, whereas some examples are monoclinic VO_2 (M), tetragonal VO_2 (R) and several metastable forms of VO_2 (A), VO_2 (B) and VO_2 (C) [4]. Among these phases, only the rutile VO_2 (R/M) phase undergoes a fully reversible metal insulator transition at a critical temperature (T_c) [1], where an abrupt alteration in optical and electronic properties is observed making it ideal for optoelectronic switches [5], memristors [6], artificial neuron networks [7,8] and intelligent window coatings [9,10].

The high temperature phase (T > T_c), has a tetragonal type structure characterized by chains of edge sharing [VO_6] octahedral along the *c*-axis with equidistant vanadium atoms (V-V = 2.88 Å) [11]. While, the low temperature structure involves V^{4+}-V^{4+} pairing with alternate shorter (0.265 nm) and longer (0.312 nm) V^{4+}-V^{4+} distances along the *a*-axis and tilting with respect to the rutile *c*-axis [11]. At 25 °C, the lattice has unit cell parameters; *a* = 5.75 Å, *b* = 4.52 Å, *c* = 5.38 Å and β = 122.60° [12]. The lattice is the result of the distortion occurring at the high temperature metallic tetragonal phase.

The mechanism of metal insulator transition in VO_2 has been investigated through computational, experimental and theoretical studies [13–15]. Nevertheless, the mechanism of the transition remains unresolved, since the VO_2 phases exhibit diverse lattice structures but have analogous electronic properties.

3. Advancements

There have been numerous studies on the VO_2 grown by APCVD since Maruyama and Ikuta utilized vanadium (III) acetylacetonate ($V(acac)_3$) as a single-precursor to deposit polycrystalline pure VO_2 films on fused quartz and sapphire single crystals [16]. In this review article, we will focus on the progress taking place during the last four years regarding the control of the processing parameters to isolate the VO_2 phases strengthening the functional properties of APCVD VO_2 layers.

The growth of amorphous pure and tungsten doped VO_2 coatings is possible on SnO_2-precoated glass substrates using vanadyl (V) triisopropoxide ($VO(OC_3H_7)_3$) as single-precursor [9,10,17,18]. It is interesting to note that the presence of tungsten in the lattice of VO_2 changed the surface morphology to worm-like (Figure 3) from granular structure [9]. This approach has several advantages including the high vapour pressure of the precursor (i.e., decomposition over time and transport of unknown species are prevented). Additionally, the operations are simplified by removing the commonly necessary oxygen source, which is usually provided either in the form of pure gas or from an extra bubbler through H_2O or alcohol. Vanadyl (IV) acetylacetonate ($VO(acac)_2$) along with propanol, ethanol and O_2 gas as oxygen sources is accomplished to grow VO_2 of different crystalline orientations [19,20]. The a-axis textured monoclinic is enhanced with propanol and ethanol, while the 022-oriented single phase V_2 is obtained with O_2 gas possessing grains (a-axis coatings) and agglomeration of grains forming rod-like structures (002-oriented phases). On controlling the oxygen gas flow rate (Figure 4), isolated monoclinic and metastable VO_2 phases can also be achieved using $VO(acac)_2$ as vanadium precursor on flexible [21] and SnO_2-precoated glass substrates [22].

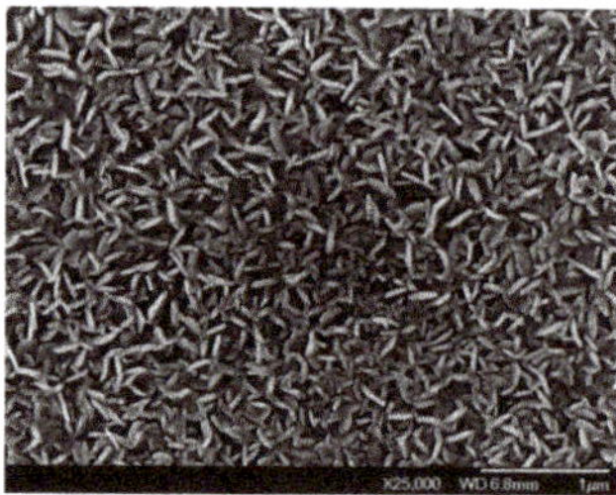

Figure 3. Field emission-scanning electron microscopy image of the APCVD tungsten doped VO_2 coating.

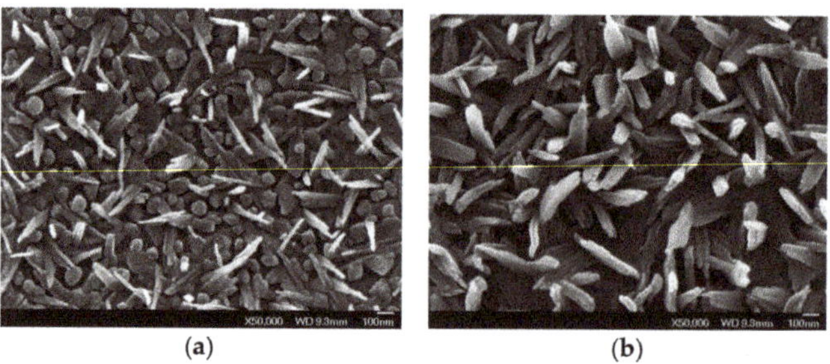

(a)   (b)

Figure 4. Field emission-scanning electron microscopy images of APCVD vanadium oxides using oxygen flow rate of 0.4 (**a**) and 0.8 L min^{-1} (**b**).

SnO$_2$ was chosen as a substrate due to the similar crystalline structure with VO$_2$, which can act as a template for the growth of rutile VO$_2$ and promote the crystallinity of the oxide [23].

4. Challenges

A comparative study among VO(acac)$_2$ and VCl$_4$, the most utilized vanadium precursors for APCVD VO$_2$, indicated that the transport rate of VO(acac)$_2$ is lower than VCl$_4$ [24]. This can be handled by increasing the temperature and the N$_2$ flow rate in the bubbler. However, this is not anticipated because the precursor may decompose over time leading to irreproducible delivery rates and the transport of unknown species. On the other hand, VCl$_4$ is highly reactive with H$_2$O resulting in inhomogeneous films [25]. A new approach uses the ethyl acetate (EtAc) as an excellent oxygen precursor resulting in the precise control of the growth rate and porosity of the films after the optimization of VCl$_4$/EtAc system [26]. A route to improve this system involves the combination of X-ray photoelectron spectroscopy (XPS) and X-ray absorption near-edge structure (XANES) to determine the effect of the substrate choice on the VO$_2$ formation for functional properties such as thermochromism [27]. It is then possible to grow VO$_2$ (Figure 5) onto substrates that induce lattice matching (SnO$_2$) or others (F-doped SnO$_2$) that promote a destabilization of V^{4+} ions and a further increase in V^{5+} deteriorating the functional properties (Figure 6).

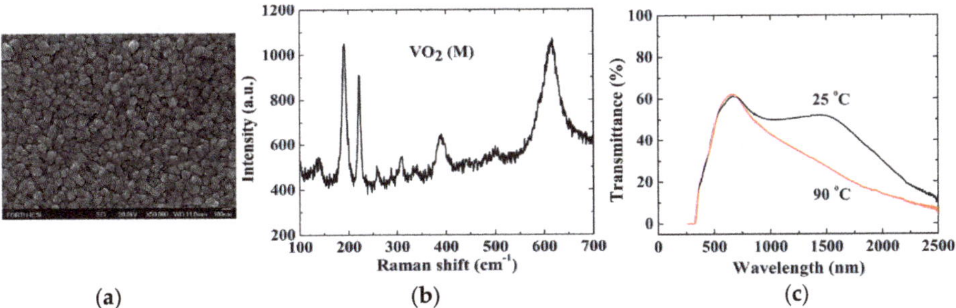

Figure 5. Field emission-scanning electron microscopy image (**a**) and (**b**) Raman spectroscopy of APCVD VO$_2$ grown on SnO$_2$-precoated glass substrates. (**c**) Transmittance spectra below T$_c$ at 25 °C and above T$_c$ at 90 °C over the region of 250–2500 nm to study the thermochromic performance.

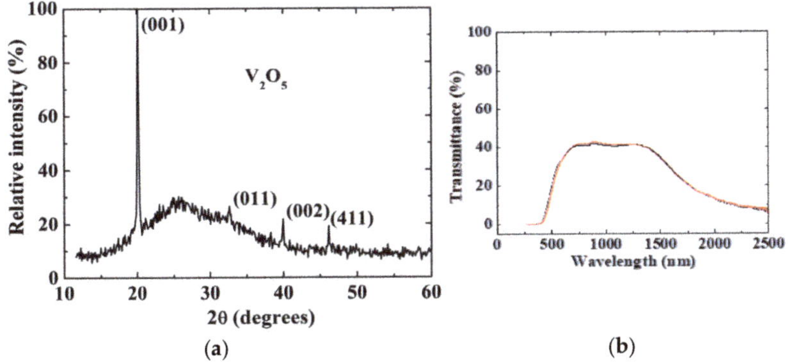

Figure 6. (**a**) X-ray diffraction of APCVD V$_2$O$_5$ grown on F-doped SnO$_2$. (**b**) No change in transmittance spectra observed over the region 250–2500 nm below T$_c$ at 25 °C (black colour) and above T$_c$ at 90 °C (red colour).

Furthermore, monoclinic VO_2 exhibits poor adhesion and is chemically susceptible to attack, restricting the use as solar control coating. In that respect, multi-functional, robust APCVD $VO_2/SiO_2/TiO_2$ films on glass substrates demonstrates excellent solar modulation properties, high transparency and resistance to abrasion compared to single VO_2 films of the same thickness [28].

5. Prospects and Outlook

In the field of APCVD VO_2, the altering of the processing parameters and the manipulation of the substrate surface is just starting to be understood. New evolvements in experimental procedures such as the utilization of single vanadium precursor and the oxygen source have addressed APCVD routes in isolating the intrinsic material properties. There are numerous exciting challenges in developing VO_2 with functional properties, which expand our understanding of the underlying chemistry and potentially lead to anticipated applications.

Two-dimensional (2D) VO_2 can also be possible by APCVD through Computational Fluid Dynamics (CFD) simulations. CFD simulations are performed to evaluate and define the whole experimental process, before, while and after the experimental procedure isolating the intrinsic material properties (Figure 7). CFD results of exhaust and quartz tube presented the simulation procedure regarding the flow rates and the temperature distribution along the boundaries of the metallic parts. The flow rate of N_2 was set at 0.1 L min^{-1} and the temperature in the inner boundaries was at 300 °C. Every aspect of the APCVD process is simulated to approach the optimal characteristics of the oxide in tandem to the surface to be deposited. Prospects in developing the growth of high-quality large-area materials with well-defined sizes, high dispersion and excellent control on layer thickness will then appear. The potential impact is illustrated by considering the exploitation possibilities of the high-performance materials by APCVD to create advanced devices for practical applications.

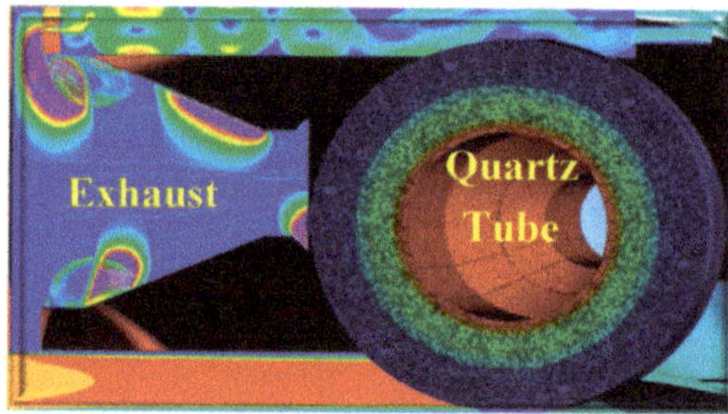

Figure 7. Simulation of transport species within the components of the APCVD reactor. The simulation showed an increase of the fluid's velocity from 2.359×10^{-3} m s^{-1} to 3.283×10^{-3} m s^{-1}, i.e., an increase of velocity of 39.18% due to temperature change. (Image courtesy of Delta Nano—Engineering Solutions Ltd., London, UK).

Acknowledgments: The authors would like to thank Pilkington Glass, UK for the supply of substrates and Aleka Manousaki for the help with the SEM characterization.

Conflicts of Interest: The authors declare no conflict of interest.

References

1. Drosos, C.; Vernardou, D. Perspectives of energy materials by APCVD. *Sol. Energy Mater. Sol. C* **2015**, *140*, 1–4. [CrossRef]
2. Chen, X.; Wang, X.; Wang, Z.; Wan, J.; Liu, J.; Qian, Y. An ethylene glycol reduction approach to metastable VO_2 nanowire arrays. *Nanotechnology* **2004**, *15*, 1685–1687. [CrossRef]
3. Graf, D.; Schläfer, J.; Garbe, S.; Klein, A.; Mathur, S. Interdepedence of structure, morphology and phase transitions in CVD grown VO_2 and V_2O_3 nanostructures. *Chem. Mater.* **2017**, *29*, 5877–5885. [CrossRef]

4.   Wang, S.; Liu, M.; Kong, L.; Long, Y.; Jiang, X.; Yu, A. Recent progress in VO_2 smart coatings: Strategies to improve the thermochromic properties. *Prog. Mater. Sci.* **2016**, *81*, 1–54. [CrossRef]

5.   Joushaghani, A.; Jeong, J.; Paradis, S.; Alain, D.; Aitchison, J.S.; Poon, J.K.S. Wavelength-size hybrid Si-VO_2 waveguide electroabsorption optical switches and photodetectors. *Opt. Express* **2015**, *23*, 3657–3688. [CrossRef] [PubMed]

6.   Driscoll, T.; Kim, H.-T.; Chae, B.-G.; Ventra, M.D.; Basov, D.N. Phase-transition driven memristive system. *Appl. Phys. Lett.* **2009**, *95*, 043503. [CrossRef]

7.   Pickett, M.D.; Medeiros-Ribeiro, G.; Williams, R.S. A scalable neuristor built with Mott memristors. *Nat. Mater.* **2013**, *12*, 114–117. [CrossRef] [PubMed]

8.   Zhou, Y.; Ramanathan, S. Memory and neuromorphic devices. *Proc. IEEE* **2015**, *103*, 1289–1310. [CrossRef]

9.   Louloudakis, D.; Vernardou, D.; Spanakis, E.; Suchea, M.; Kenanakis, G.; Pemble, M.E.; Savvakis, C.; Katsarakis, N.; Koudoumas, E.; Kiriakidis, G. Atmospheric pressure chemical vapour deposition of amorphous tungsten doped vanadium dioxide for smart window applications. *Adv. Mater. Lett.* **2016**, *7*, 192–196. [CrossRef]

10. Vernardou, D.; Louloudakis, D.; Spanakis, E.; Katsarakis, N.; Koudoumas, E. Amorphous thermochromic VO_2 coatings grown by APCVD at low temperatures. *Adv. Mater. Lett.* **2015**, *6*, 660–663. [CrossRef]

11. Morrison, V.R.; Chatelain, R.P.; Tiwari, K.L.; Hendaoui, A.; Bruhács, A.; Chaker, M.; Siwick, B.J. A photoinduced metal-like phase of monoclinic VO_2 revealed by ultrafast electron diffraction. *Science* **2014**, *346*, 445–448. [CrossRef] [PubMed]

12. Eyert, V. The metal-insulator transition of VO_2: A band theoretical approach. *Ann. Phys.* **2002**, *11*, 650–704. [CrossRef]

13. Warwick, M.E.A.; Binions, R. Advances in thermochromic vanadium dioxide films. *J. Mater. Chem. A* **2014**, *2*, 3275–3292. [CrossRef]

14. Goodenough, J.B. The two components of the crystallographic transition in VO_2. *J. Solid State Chem.* **1971**, *3*, 490–500. [CrossRef]

15. Wentzcovitch, R.M.; Schulz, W.W.; Allen, P.B. VO_2: Peierls or Mott-Hubbard? A view from band theory. *Phys. Rev. Lett.* **1994**, *72*, 3389–3392. [CrossRef] [PubMed]

16. Maruyama, T.; Ikuta, Y. Vanadium dioxide thin films prepared by chemical vapour deposition from vanadium(III) acetylacetonate. *J. Mater. Sci.* **1993**, *28*, 5073–5078. [CrossRef]

17. Louloudakis, D.; Vernardou, D.; Spanakis, E.; Katsarakis, N.; Koudoumas, E. Thermochromic vanadium oxide coatings grown by APCVD at low temperatures. *Phys. Procedia* **2013**, *46*, 137–141. [CrossRef]

18. Vernardou, D.; Louloudakis, D.; Spanakis, E.; Katsarakis, N.; Koudoumas, E. Thermochromic amorphous VO_2 coatings grown by APCVD using a single-precursor. *Sol. Energy Mater. Sol. C* **2014**, *128*, 36–40. [CrossRef]

19. Vernardou, D.; Bei, A.; Louloudakis, D.; Katsarakis, N.; Koudoumas, E. Oxygen source-oriented control of atmospheric pressure chemical vapour deposition of VO_2 for capacitive applications. *J. Electrochem. Sci. Eng.* **2016**, *6*, 165–173. [CrossRef]

20. Louloudakis, D.; Vernardou, D.; Spanakis, E.; Dokianakis, S.; Panagopoulou, M.; Raptis, G.; Aperathitis, E.; Kiriakidis, G.; Katsarakis, N.; Koudoumas, E. Effect of O_2 flow rate on the thermochromic performance of VO_2 coatings grown by atmospheric pressure CVD. *Phys. Status Solidi C* **2015**, *12*, 856–860. [CrossRef]

21. Vernardou, D.; Louloudakis, D.; Spanakis, E.; Katsarakis, N.; Koudoumas, E. Functional properties of APCVD VO_2 layers. *Int. J. Thin Films Sci. Technol.* **2015**, *4*, 187–191. [CrossRef]

22. Vernardou, D.; Apostolopoulou, M.; Louloudakis, D.; Katsarakis, N.; Koudoumas, E. Electrochemical performance of vanadium oxide coatings grown using atmospheric pressure CVD. *Chem. Vap. Depos.* **2015**, *21*, 369–374. [CrossRef]

23. Zhang, Z.; Gao, Y.; Luo, H.; Kang, L.; Chen, Z.; Du, J.; Kanehira, M.; Zhang, Y.; Wang, Z.L. Solution-based fabrication of vanadium dioxide on $F:SnO_2$ substrates with largely enhanced thermochromism and low-emissivity for energy-saving applications. *Energy Environ. Sci.* **2011**, *4*, 4290–4297. [CrossRef]

24. Vernardou, D.; Pemble, M.E.; Sheel, D.W. The growth of thermochromic VO_2 films on glass by atmospheric-pressure CVD: A comparative study of precursors, CVD methodology and substrates. *Chem. Vap. Depos.* **2006**, *12*, 263–274. [CrossRef]

25. Béteille, F.; Mazerolles, L.; Livage, J. Microstructure and metal-insulating transition of VO_2 thin films. *Mater. Res. Bull.* **1999**, *34*, 2177–2184. [CrossRef]

26. Malarde, D.; Powell, M.J.; Quesada-Cabrera, R.; Wilson, R.L.; Carmalt, C.J.; Samkar, G.; Parkin, I.P.; Palgrave, R.G. Optimized atmospheric-pressure chemical vapour deposition thermochromic VO_2 thin films for intelligent window applications. *ACS Omega* **2017**, *2*, 1040–1046. [CrossRef]

27. Powell, M.J.; Godfrey, I.J.; Quesada-Cabrera, R.; Malarde, D.; Teixeira, D.; Emerich, H.; Palgrave, R.G.; Carmalt, C.J.; Parkin, I.P.; Sankar, G. Qualitative XANES and XPS analysis of substrate effects in VO_2 thin films: A route to improving chemical vapour deposition synthetic methods? *J. Phys. Chem. C.* **2017**, *121*, 20345–20352. [CrossRef]

28. Powell, M.J.; Quesada-Cabrera, R.; Taylor, A.; Teixeira, D.; Papakonstantinou, I.; Palgrave, R.G.; Sankara, G.; Parkin, I.P. Intelligent multifunctional $VO_2/SiO_2/TiO_2$ coatings for self-cleaning, energy-saving window panels. *Chem. Mater.* **2016**, *28*, 1369–1376. [CrossRef]

Letter

Electrochromic Performance of V_2O_5 Thin Films Grown by Spray Pyrolysis

Kyriakos Mouratis [1,2,*], **Valentin Tudose** [1,3], **Cosmin Romanitan** [4], **Cristina Pachiu** [4], **Oana Tutunaru** [4], **Mirela Suchea** [1,4,*], **Stelios Couris** [2], **Dimitra Vernardou** [1,5] and **Koudoumas Emmanouel** [1,5,*]

[1]   Center of Materials Technology and Photonics, School of Engineering, Hellenic Mediterranean University, 71410 Heraklion, Greece; tudose_valentin@yahoo.com (V.T.); dvernardou@hmu.gr (D.V.)

[2]   Department of Physics, University of Patras, 26500 Patras, Greece; couris@physics.upatras.gr

[3]   Department of Chemistry, University of Crete, 70013 Heraklion, Greece

[4]   National Institute for Research and Development in Microtechnologies—IMT Bucharest, 126A, Erou Iancu 8 Nicolae Street, 077190 Bucharest, Romania; cosmin.romanitan@imt.ro (C.R.); cristina.pachiu@gmail.com (C.P.); oana.tutunaru@imt.ro (O.T.)

[5]   Department of Electrical and Computer Engineering, School of Engineering, Hellenic Mediterranean University, 71410 Heraklion, Greece

*   Correspondence: kmuratis@hmu.gr (K.M.); mirasuchea@hmu.gr (M.S.); koudoumas@hmu.gr (K.E.)

Received: 5 August 2020; Accepted: 29 August 2020; Published: 1 September 2020

Abstract: A new approach regarding the development of nanostructured V_2O_5 electrochromic thin films at low temperature (250 °C), using air-carrier spray deposition and ammonium metavanadate in water as precursor is presented. The obtained V_2O_5 films were characterized by X-ray diffraction, scanning electron microscopy and Raman spectroscopy, while their electrochromic response was studied using UV-vis absorption spectroscopy and cyclic voltammetry. The study showed that this simple, cost effective, suitable for large area deposition method can lead to V_2O_5 films with large active surface for electrochromic applications.

Keywords: vanadium pentoxide; electrochromic; spray pyrolysis; ammonium metavanadate

---

1. Introduction

Currently, many efforts are going on worldwide to develop new electrochromic materials for energy and environmental applications. V_2O_5 compounds keep attracting much attention because of their structural flexibility and chemical/physical properties, which are suitable for catalytic and electrochemical applications [1–3]. The layered structure with orthorhombic symmetry of α-V_2O_5 can be well adapted to the reversible incorporation of guest Li^+ ions. As a result, its films have been recognized as the most attractive material for applications in electrochromic devices and in lithium batteries [3–11]. The valence state of V atoms is known to be extremely sensitive to the chemical environment, a behavior that can lead to a variety of structures containing different building units and exhibiting different physical properties. In particular, the structure and the morphology of vanadium oxide films are intimately related to the deposition method and the operating conditions. Moreover, since the direct growth of crystalline V_2O_5 films is very difficult except in the cases of some sub-stoichiometric oxides, annealing of the as-prepared films might be required in order to improve properties and functionality. Over the years, many techniques were involved for the deposition of V_2O_5 thin films for electrochromic applications, including electrodeposition, sputtering, sol-gel, hydrothermal growth and doctor-blade [12–16]. Some trials have been also reported on the use of spray pyrolysis for the deposition of large surfaces electrochromic V_2O_5 films, most of them using relatively high temperatures (>500 °C) vanadium chloride (i.e., moisture sensitive and toxic), vanadium nitride (i.e., irritant) as precursors [17–20].

The present letter concerns preliminary results on a new approach for developing nanostructured V_2O_5 thin films for electrochromic applications. In particular, air-carrier spray deposition at quite low temperature has been employed with ammonium metavanadate dissolved in water, which is a rather simple and low-cost precursor.

2. Materials and Methods

2.1. Materials

Ammonium metavanadate, ACS reagent, ≥99.0% (NH_4VO_3) was purchased from Sigma-Aldrich (St. Louis, MO, USA) and distilled water was used as a solvent for the precursor's preparation. Lithium perchlorate, ACS reagent, ≥95.0% ($LiCLO_4$) from Sigma-Aldrich was used as electrolyte and propylene carbonate, anhydrous, 99.7% from Sigma-Aldrich was the electrolyte's solvent. Finally, fluorine-doped tin Oxide ($FTO–SnO_2$:F) coated glass was the substrate.

2.2. Preparation of V_2O_5 Samples

A custom-made spray pyrolysis technique was utilized for the preparation of the samples. The deposition was carried out at a distance of 30 cm between the spray gun and the substrate, which was kept at a temperature of 250 °C. The precursor solution was prepared by dissolving the required amount of ammonium metavanadate (NH_4VO_3) in distilled water. Two concentrations were used, 0.01 M and 0.02 M. Since the optical difference between the colored and the bleached states of the as deposited samples was very low as well as their electrochromic performance was quite poor, thermal annealing was carried out for 2 h at 400 °C in oxygen atmosphere. More details regarding the sample names and the growth parameters can be found in Table 1. The respective thickness was measured to be about a micron.

Table 1. Growth parameters of V_2O_5 thin films.

Sample Name	Substrate	Precursor Volume (mL)	Temperature (°C)	Concentration (M)	Annealing
SV1	FTO	20	250	0.02	No
SV2	FTO	20	250	0.01	No
SV1_TT	FTO	20	250	0.02	400 °C for 2 h
SV2_TT	FTO	20	250	0.01	400 °C for 2 h

2.3. Structural and Morphological Characterization

X-ray diffraction (XRD), scanning electron microscopy (SEM) and Raman Spectroscopy were used to analyze the structure and the morphology of the deposited V_2O_5 films. For the XRD measurements, a 9kW SmartLab X-ray Diffraction System (Rigaku, Tokyo, Japan) was employed, with rotating anode that employs Cu Kα1 radiation (λ = 1.5406 Å) in parallel-beam mode. To reveal the crystalline structure of the layer, the grazing incidence technique was used, the incidence angle was kept at 0.5°, while the detector was moved from 5 to 95°. SEM characterization was performed using a Nova NanoSEM 630 (FEI Company, Hillsborough, OR, USA), field emission scanning electron microscope for the annealed samples so that a better resolution and insight on their fine surface structuring can be obtained. All samples were characterized in high vacuum mode without any coating. Raman spectroscopy was performed at room temperature using a WiTec Raman spectrometer (Alpha-SNOM 300 S, WiTec GmbH, Ulm, Germany) using 532 nm excitation, from a diode-pumped solid-state laser with a maximum power of 145 mW. The incident laser beam with a spot-size of about 1.0 μm was focused onto the sample with a 100× long working distance microscope objective. The Raman spectra were collected with an exposure time of 20 s accumulation and the scattered light was collected by the same objective in back-scattering geometry using a 600 grooves/mm grating. The spectrometer scanning data collection and processing were carried out using the WiTec Project Five software (Version, WiTec GmbH, Ulm, Germany).

2.4. Electrochemical Characterization

Electrochemical experiments were performed using a PGSTAT302N Autolab (Metrohm AG, Herisau, Switzerland) potentiostat/galvanostat in a three-electrode cell setup, the working electrode with the V_2O_5 layer on the FTO substrate, the Pt counter electrode and the reference electrode (Ag/AgCl) utilizing 1M, $LiClO_4$ in propylene carbonate as electrolyte [21].

2.5. Optical Characterization

The optical properties of the samples were evaluated by UV-Vis transmittance spectra recorded with a UV-2401PC spectrophotometer (Shimadzu Corporation, Kyoto, Japan,), at a wavelength range of 200 nm to 1000 nm.

3. Results and Discussion

3.1. Morphological Characterization

As shown in Figure 1a,b, SEM characterization revealed the formation of mixed nanostructured granular and wall-like structures films, homogeneously distributed over the film surface.

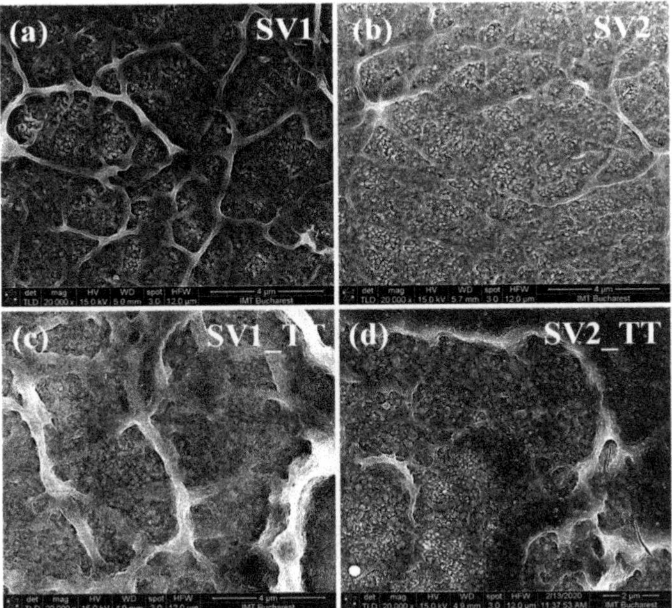

Figure 1. (**a**) SEM image of SV1 sample, (**b**) SEM image of SV2 sample, (**c**) SEM image of SV1_TT sample, (**d**) SEM image of SV2_TT sample.

Moreover, as can be observed, the films present a very large active surface when compared with usual flat granular structured films. Annealing of the films for 2 h at 400 °C in an oxygen atmosphere was found to lead to a decrease of the number of film surface features, as can be seen in Figure 1c,d. This kind of structuring has not been reported before when using other growth techniques and it seems to be directly related to the growth technique used in our work. The particular surface structuring and morphology makes these films suitable for applications that require a high surface to volume ratio.

3.2. X-ray Diffraction Analysis

Grazing incidence X-ray diffraction (GIXRD) was used to reveal the crystalline structure of the investigated samples. Since GIXRD requires small incident angles for the incoming X-rays, the diffraction is surface sensitive, the wave penetration being limited, and this approach is used for the study surfaces and layers. Figure 2a shows the GIXRD patterns of SV1 and SV2 before (red line) and after annealing (blue line).

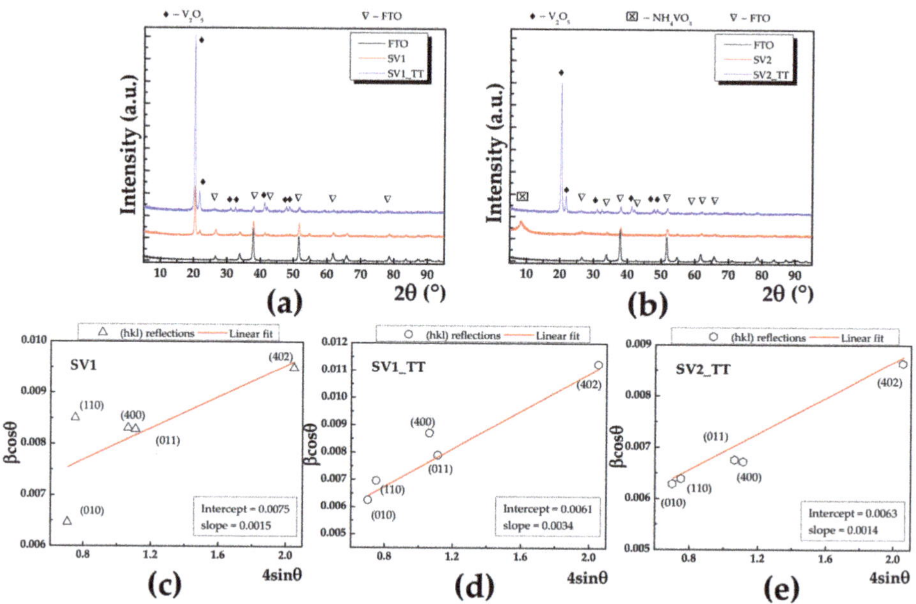

Figure 2. GIXRD patterns for (**a**) SV1 before and after annealing, (**b**) SV2 before and after annealing with red and blue line, respectively. In addition, GIXRD pattern corresponding to FTO was added (black line). (**c–e**) Williamson-Hall plots for SV1, SV1_TT and SV2, respectively with corresponding values for intercept and slope as inset.

In addition, the corresponding GIXRD pattern of the FTO substrate was also added (black line). The peak indexing was made using International Center for Diffraction Database (ICDD) database. In the case of SV1, the presence of the V_2O_5 phase was identified, with orthorhombic lattice having parameters a = 11.7 Å, b = 4.40 Å and c = 3.45 Å, while $\alpha = \beta = \gamma = 90°$, which belongs to the 47:Pmmm space group. To gain a thoroughly view of the crystalline features of our materials, we employed Williamson-Hall method, which allowed us to separate the strain and size effects from the Bragg peak. The formalism behind this method can be found in [22]. Figure 2c–e present the Williamson-Hall plots (red line) for SV1, SV1_TT and SV2_TT with corresponding intercept and slope values as inset, which are related to the mean crystallite size and lattice strain, respectively. As a result, the mean crystallite size was increasing from 18.4 to 22.7 nm. For the respective as deposited SV2 sample, the XRD pattern exhibited only a broad peak at 8.07°, which can be attributed to the non-reacted precursor NH_4VO_3 since it cannot be matched with any vanadium oxide phase. After annealing, the occurrence of V_2O_5 was visible and the respective mean crystallite size calculated to be 22.1 nm. At the same time, after the annealing process, the lattice strain ε increases from 0.15% (e.g., SV1) to 0.34% (e.g., SV1_TT), which indicates a slight expansion of the interplanar distance. The XRD results indicate that annealing can result in two main effects.

Firstly, the crystallinity is enhanced, which leads to a smaller dislocation density, whose occurrence is due to the boundaries of the adjacent mosaic blocks [23]. Simultaneously, the decreasing of dislocation density is accompanied by a slight expansion of the lattice strain. Secondly, the full conversion from NH_4VO_3 to V_2O_5 phase can take place.

3.3. Raman Analysis

Since the structure of α-V_2O_5 (orthorhombic) belongs to the Pmmn space group [24,25], the V_2O_5 layers are formed from packing of edge shared VO_5 square pyramids linked in the 'XY' plane. Group theoretical analysis predicts twenty-one Raman active modes for V_2O_5 at Γ point, 7Ag + 7B2g + 3B1g + 4B3g [26,27]. In our case we observed five Raman modes for the annealed films as shown in Figure 3, which match with the reported data for V_2O_5.

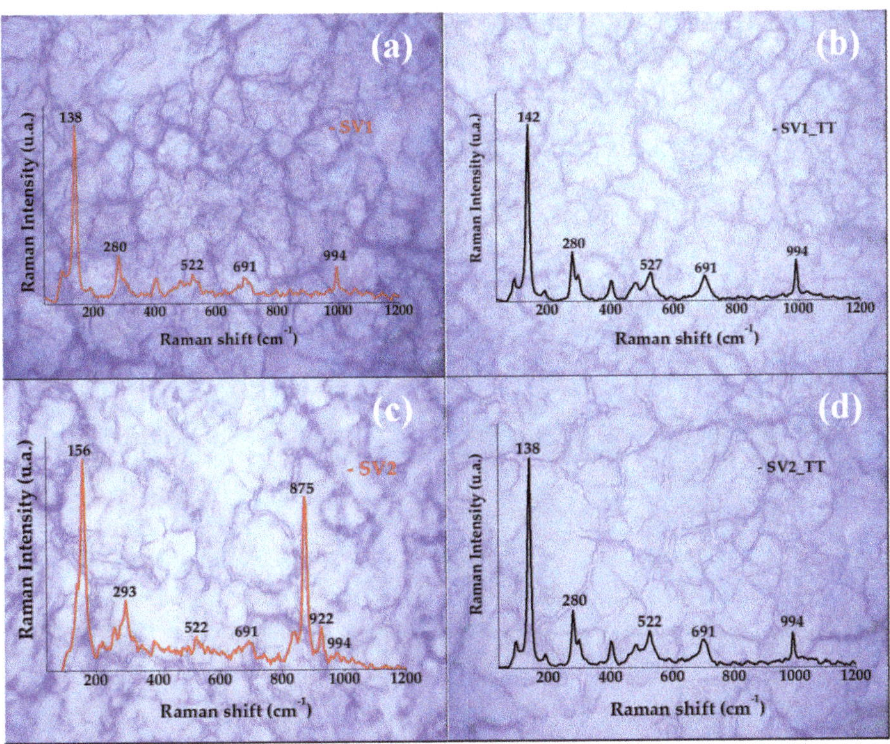

Figure 3. Raman spectra overlapped over optical microscopy images (100× magnification) of SV1 (20 mL, 0.02 M) and SV2 (20 mL, 0.01 M) samples, (**a**,**c**) before annealing, (**b**,**d**) after annealing.

In particular, the Raman peaks we observed were at 138, 280, 522, 691 and 994 cm^{-1}, confirming the presence of α-V_2O_5 phase [26], as was already identified with XRD. The 5 observed Raman peaks can be assigned as follows:

(a) The highest frequency peak at 994 cm^{-1} appears to be due to the stretching vibrational mode of V–OI bond along Z direction.

(b) Displacement of OIII atoms in Y and X directions generates Raman modes at 691 cm^{-1} (V–OIII–V antiphase stretching mode) and 522 cm^{-1} (d4 stretching vibration), respectively.

(c)   Mode 280 cm^{-1} can be attributed to oscillation of OI atoms along Y axis.

(d)   The low frequency mode at 138 cm^{-1} corresponds to Y displacement of the whole chain involving shear and rotations of the ladder like V–OIII bonds. The high intensity of 138 cm^{-1} peak indicates the long range order of V–O layers in the XY plane.

Considering the annealing effect on the film structuring, the most affected modes by the temperature were found to be the V^{+2}=O stretching bands situated at 138 cm^{-1} which has a slight deviation to the left and the V^{+3}=O situated at 522 cm^{-1} which increases in Raman intensity. The other V$_2$O$_5$ vibration modes (V^{+3}=O stretching modes situated at 691 cm^{-1} and V^{+5}=O stretching modes situated at 994 cm^{-1}) were not affected by the compressive stress introduced by temperature [28]. Supplementary peaks associated to nonreacted metavanadate precursor, were detected at 871 cm^{-1} in the SV2 current study.

3.4. Electrochemical Characterization

The electrochemical performance of the samples was studied in a three-electrode cell, by cycling the potential between −1 V and +1 V at a scan rate of 10 mV s^{-1} and using 1M, LiClO$_4$ in propylene carbonate as electrolyte (Figure 4a).

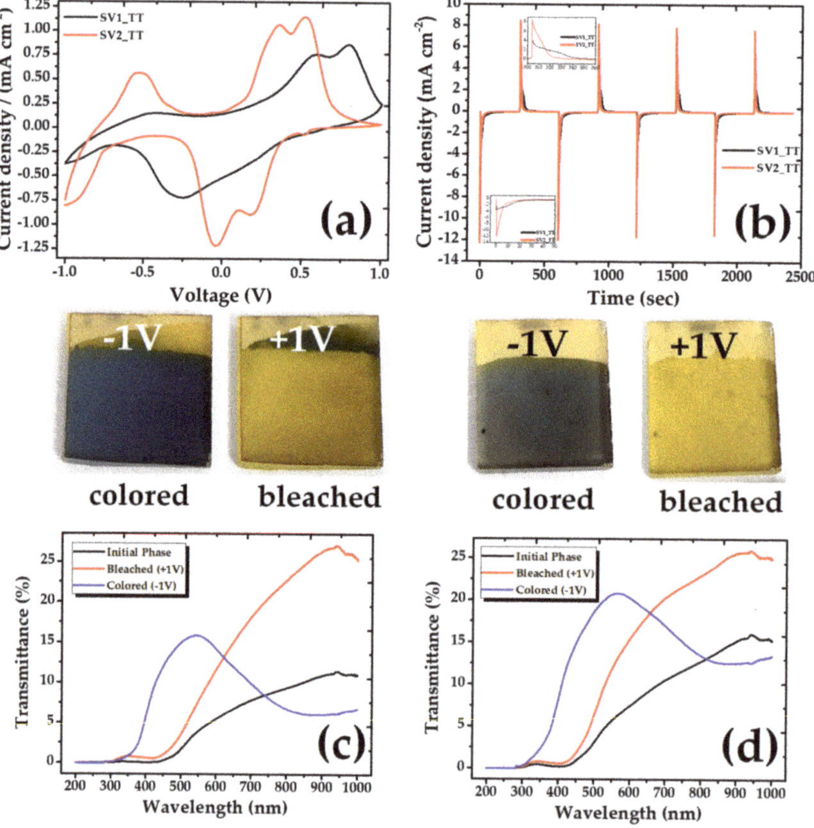

Figure 4. (**a**) Cyclic voltammograms after annealing, (**b**) Chronoamperometry after annealing, (**c,d**) UV-Vis transmittance comparison of the initial, bleached and colored state of the V$_2$O$_5$ coatings, (**c**) SV1_TT, (**d**) SV2_TT.

In parallel, their electrochromic performance was studied with UV-Vis absorption spectroscopy. A minimal optical difference between colored and bleached states of the as deposited samples along with poor electrochromic properties were indicated. In particular, the color varied between light gray/yellow (−1 V) and light yellow (+1 V). In order to improve the structure of the V_2O_5 films and their subsequent performance, the samples were annealed by heating them in an oven at a temperature of 400 °C for 2 h in an oxygen atmosphere. After annealing, the color states changed and improved dramatically, being blue for the SV1 sample and a darker gray for the respective SV2, as compared to the original colors before the annealing process. Comparing the bleached states, the annealed samples have stronger yellow tint than the samples before the annealing. The color changes of the annealed samples are presented in Figure 4.

The SV2_TT coating indicates three cathodic peaks at −0.926 V/−0.010 V/+0.230 V and three anodic peaks at −0.634 V/+0.328 V/+0.485 V, which are attributed to Li^+ intercalation and deintercalation accompanying gain and loss of an e^-. The current density of the particular sample is the highest indicating an enhanced electrochemical activity. On the other hand, the curve of SV1_TT is different showing two anodic and one cathodic peak (Figure 4a). This discrepancy may be related to the expansion of the interplanar distance of SV2_TT, which allows accommodating a greater number of Li ions.

Chronoamperometry measurements were used to evaluate the time evolution of the coloring/ bleaching processes (Figure 4b). When a negative voltage (−1.0 V) was applied, the color of the V_2O_5 coating changed from yellow to blue (SV1_TT sample) or dark grey (SV2_TT sample). With the opposite applied voltage (+1.0 V), the film returned to its initial state, becoming yellow again. From the I–t measurements, the time responses for coloration of the SV1_TT and SV2_TT samples were found to be 34.8 s and 9.8 s, respectively, the respective times for bleaching being 34.2 s and 15.7 s. As can be observed, SV2_TT has a much faster response. The charge densities of these samples, calculated by integrating the current density, were found to be −73.3/−71.01 and 70.7/70.9 mC cm^{-2} for the SV1_TT and SV2_TT samples, respectively.

The transmittance measurements performed in the annealed samples are also shown in Figure 4c, d. In these graphs, we can observe how the transmittance spectra change between the colored and the bleached state according to the applied voltages in the samples. More specifically, it is obvious that above 650 nm, the color difference is increasing, a behavior indicating that the electrochromic performance of our samples is maximized at the near infrared spectral region.

Finally, the coloration efficiency (*CE*) factor, defined as the change in optical density (ΔOD) per unit of charge density (*Q*), was calculated according to the following Equation (1):

$$CE(\lambda) = \frac{\Delta OD(\lambda)}{Q} \tag{1}$$

where ΔOD is the optical density, showing the change in the transmission between the bleached and colored states of the film, that is, Equation (2):

$$\Delta OD(\lambda) = \ln\left(\frac{T_b}{T_c}\right) \tag{2}$$

where, T_b and T_c are the transmittances at bleached and colored state, respectively. Q is the charge density and λ denotes certain wavelengths, which, for our calculations, were taken at 700 and 900 nm. The *CE* for our films was calculated to be 8.1 and 2.4 cm^{-2} C^{-1} for SV1_TT and SV2_TT respectively at 700 nm, increasing up to 20 and 10 cm^{-2} C^{-1} at 900 nm. Although the coloration efficiency is higher for the SV1_TT sample, the experiments showed that this is decaying faster than the SV2_TT one, which is presenting better crystalline properties. Since these are preliminary result, further efforts are currently under way in order to increase both the coloration efficiency and the stability of the films.

4. Conclusions

Electrochromic nanostructured V_2O_5 thin films were prepared at low temperature (250 °C) using air-carrier spray deposition, starting from ammonium metavanadate precursor in water. The obtained V_2O_5 films were characterized by X-ray diffraction, scanning electron microscopy and Raman spectroscopy, while their electrochromic behavior was studied using UV-vis absorption spectroscopy and cyclic voltammetry. This preliminary study showed that this simple, cost effective, suitable for large area deposition method can lead to novel surface structuring of V_2O_5 films with electrochromic performance. Further studies for growth optimization and improvements of film properties and stability would be performed.

Author Contributions: Conceptualization, K.E., S.C. and M.S.; methodology, V.T., M.S. and D.V.; validation, K.M., K.E. and M.S.; formal analysis, K.M., V.T. and M.S.; investigation, C.R., O.T., C.P. and K.M.; data curation, K.M., V.T., M.S. and D.V.; writing—original draft preparation K.M., C.R., C.P. and M.S.; writing—review and editing K.M., M.S. and K.E.; supervision, K.E., S.C. and M.S.; project administration, K.E.; funding acquisition, K.M. and K.E. All authors have read and agreed to the published version of the manuscript.

Funding: This research is co-financed by Greece and the European Union (European Social Fund—ESF) through the Operational Program "Human Resources Development, Education and Lifelong Learning" in the context of the project "Strengthening Human Resources Research Potential via Doctorate Research—2nd Cycle" (MIS-5000432), implemented by the State Scholarships Foundation (IKY).

Acknowledgments: M.S. involvement was partially supported by the Romanian Ministry of Research and Innovation through Institutional Performance—Projects for Excellence Financing in RDI, Contract no. 13 PFE/ 16.10.2018.

Conflicts of Interest: The authors declare no conflict of interest.

References

1. Weckhuysen, B.M.; Keller, D.E. Chemistry, Spectroscopy and the Role of Supported Vanadium Oxides in Heterogeneous Catalysis. *Catal. Today* **2003**, *78*, 25–46. [CrossRef]
2. Chernova, N.A.; Roppolo, M.; Dillon, A.C.; Whittingham, M.S. Layered Vanadium and Molybdenum Oxides: Batteries and Electrochromics. *J. Mater. Chem.* **2009**, *19*, 2526–2552. [CrossRef]
3. Delmas, C.; Cognac-Auradou, H.; Cocciantelli, J.M.; Ménétrier, M.; Doumerc, J.P. The $Li_xV_2O_5$ System: An Overview of the Structure Modifications Induced by the Lithium Intercalation. *Solid State Ion.* **1994**, *69*, 257–264. [CrossRef]
4. Whittingham, M.S. Lithium Batteries and Cathode Materials. *Chem. Rev.* **2004**, *104*, 4271–4302. [CrossRef]
5. Whittingham, M.S.; Savinell, R.F.; Zawodzinski, T. Introduction: Batteries and Fuel Cells. *Chem. Rev.* **2004**, *104*, 4243–4244. [CrossRef]
6. Murphy, D.W.; Christian, P.A.; DiSalvo, F.J.; Waszczak, J.V. Lithium Incorporation by Vanadium Pentoxide. *Inorg. Chem.* **1979**, *18*, 2800–2803. [CrossRef]
7. Whittingham, M.S. The Role of Ternary Phases in Cathode Reactions. *J. Electrochem. Soc.* **1976**, *123*, 315. [CrossRef]
8. Wiesener, K.; Schneider, W.; Ilić, D.; Steger, E.; Hallmeier, K.H.; Brackmann, E. Vanadium Oxides in Electrodes for Rechargeable Lithium Cells. *J. Power Sources* **1987**, *20*, 157–164. [CrossRef]
9. Mukherjee, A.; Ardakani, H.A.; Yi, T.; Cabana, J.; Shahbazian-Yassar, R.; Klie, R.F. Direct Characterization of the Li Intercalation Mechanism into α-V_2O_5 Nanowires Using *in-Situ* Transmission Electron Microscopy. *Appl. Phys. Lett.* **2017**, *110*, 213903. [CrossRef]
10. Li, H.; He, P.; Wang, Y.; Hosono, E.; Zhou, H. High-Surface Vanadium Oxides with Large Capacities for Lithium-Ion Batteries: From Hydrated Aerogel to Nanocrystalline $VO_2(B)$, V_6O_{13} and V_2O_5. *J. Mater. Chem.* **2011**, *21*, 10999–11009. [CrossRef]
11. Singh, B.; Gupta, M.K.; Mishra, S.K.; Mittal, R.; Sastry, P.U.; Rols, S.; Chaplot, S.L. Anomalous Lattice Behavior of Vanadium Pentaoxide (V_2O_5): X-ray Diffraction, Inelastic Neutron Scattering and Ab Initio Lattice Dynamics. *Phys. Chem. Chem. Phys.* **2017**, *19*, 17967–17984. [CrossRef] [PubMed]
12. Beke, S. A Review of the Growth of V_2O_5 Films from 1885 to 2010. *Thin Solid Films* **2011**, *519*, 1761–1771. [CrossRef]

13. Granqvist, C.G.; Arvizu, M.A.; Bayrak Pehlivan, I.; Qu, H.-Y.; Wen, R.-T.; Niklasson, G.A. Electrochromic materials and devices for energy efficiency and human comfort in buildings: A critical review. *Electrochim. Acta* **2018**, *259*, 1170–1182. [CrossRef]

14. Panagopoulou, M.; Vernardou, D.; Koudoumas, E.; Tsoukalas, D.; Raptis, Y.S. Oxygen and temperature effects on the electrochemical and electrochromic properties of rf-sputtered V_2O_5 thin films. *Electrochim. Acta* **2017**, *232*, 54–63. [CrossRef]

15. Mjejri, I.; Manceriu, L.M.; Gaudonc, M.; Rougier, A.; Sediri, F. Nano-vanadium pentoxide films for electrochromic displays. *Solid State Ion* **2016**, *292*, 8–14. [CrossRef]

16. Chu, J.; Kong, Z.; Lu, D.; Zhang, W.; Wang, X.; Yu, Y.; Li, S.; Wang, X.; Xiong, S.; Ma, J. Hydrothermal synthesis of vanadium oxide nanorods and their electrochromic performance. *Mater. Lett.* **2016**, *166*, 179–182. [CrossRef]

17. Varadaraajan, V.; Satishkumar, B.C.; Nanda, J.; Mohanty, P. Direct Synthesis of Nanostructured V_2O_5 Films Using Solution Plasma Spray Approach for Lithium Battery Applications. *J. Power Sources* **2011**, *196*, 10704–10711. [CrossRef]

18. Mrigal, A.; Addou, M.; El Jouad, M.; Hssein, M.; Khannyra, S. Electrochemical Stability and Large Optical Modulation of the V_2O_5 Thin Films Made by Spray Pyrolysis. *Sens. Lett.* **2018**, *16*, 204–210. [CrossRef]

19. Irani, R.; Rozati, S.M.; Beke, S. Structural and Optical Properties of Nanostructural V_2O_5 Thin Films Deposited by Spray Pyrolysis Technique: Effect of the Substrate Temperature. *Mater. Chem. Phys.* **2013**, *139*, 489–493. [CrossRef]

20. Wei, Y.; Li, M.; Zheng, J.; Xu, C. Structural Characterization and Electrical and Optical Properties of V_2O_5 Films Prepared via Ultrasonic Spraying. *Thin Solid Films* **2013**, *534*, 446–451. [CrossRef]

21. Vernardou, D.; Apostolopoulou, M.; Louloudakis, D.; Katsarakis, N.; Koudoumas, E. Hydrothermally grown β-V_2O_5 electrode at 95 °C. *J. Colloid Interface Sci.* **2014**, *424*, 1–6. [CrossRef] [PubMed]

22. Williamson, G.K.; Hall, W.H. X-ray line broadening from filled aluminium and wolfram. *Acta Metall.* **1953**, *1*, 22–31. [CrossRef]

23. Williamson, G.K.; Smallman, R.E. III. Dislocation Densities in Some Annealed and Cold-Worked Metals from Measurements on the X-ray Debye-Scherrer Spectrum. *Philos. Mag. J. Theor. Exp. Appl. Phys.* **1956**, *1*, 34–46. [CrossRef]

24. Enjalbert, R.; Galy, J. A Refinement of the Structure of V_2O_5. *Acta Crystallogr. C* **1986**, *42*, 1467–1469. [CrossRef]

25. Bachmann, H.G.; Ahmed, F.R.; Barnes, W.H. The Crystal Structure of Vanadium Pentoxide. *Z. Krist.-Cryst. Mater.* **1961**, *115*, 110–131. [CrossRef]

26. Abello, L.; Husson, E.; Repelin, Y.; Lucazeau, G. Vibrational Spectra and Valence Force Field of Crystalline V_2O_5. *Spectrochim. Acta Part Mol. Spectrosc.* **1983**, *39*, 641–651. [CrossRef]

27. Baddour-Hadjean, R.; Pereira-Ramos, J.P.; Navone, C.; Smirnov, M. Raman Microspectrometry Study of Electrochemical Lithium Intercalation into Sputtered Crystalline V_2O_5 Thin Films. *Chem. Mater.* **2008**, *20*, 1916–1923. [CrossRef]

28. Abd-Alghafour, N.M.; Ahmed, N.M.; Hassan, Z. Fabrication and Characterization of V_2O_5 Nanorods Based Metal–Semiconductor–Metal Photodetector. *Sens. Actuators Phys.* **2016**, *250*, 250–257. [CrossRef]

MDPI

St. Alban-Anlage 66

4052 Basel

Switzerland

Tel. +41 61 683 77 34

Fax +41 61 302 89 18

www.mdpi.com

Materials Editorial Office

E-mail: materials@mdpi.com

www.mdpi.com/journal/materials